FOREST TOURISM AND RECREATION

Case Studies in Environmental Management

Edited by

Xavier Font and John Tribe
Faculty of Leisure and Tourism
Buckinghamshire Chilterns University College
High Wycombe
UK

CABI *Publishing*

CABI *Publishing* is a division of CAB *International*

CABI Publishing
CAB International
Wallingford
Oxon OX10 8DE
UK

Tel: +44 (0)1491 832111
Fax: +44 (0)1491 833508
Email: cabi@cabi.org

CABI Publishing
10 E 40th Street
Suite 3203
New York, NY 10016
USA

Tel: +1 212 481 7018
Fax: +1 212 686 7993
Email: cabi-nao@cabi.org

A catalogue record for this book is available from the British Library, London, UK.

Library of Congress Cataloging-in-Publication Data
Forest tourism and recreation : case studies in environmental management / edited by Xavier Font and John Tribe.
p. cm.
Includes bibliographical references.
ISBN 0-85199-414-8 (alk. paper)
1. Ecotourism. 2. Forest reserves--Recreational use. I. Font, Xavier. II. Tribe, John.
G156.5.E26F67 1999 99–41615
333.78'4—dc21 CIP

First published 2000
Reprinted 2002

This publication was produced as part of TOURFOR, a European Commission LIFE project.

ISBN 0 85199 414 8

Typeset by Columns Design Ltd, Reading.
Printed and bound in Great Britain by Biddles Ltd, *www.biddles.co.uk*

Contents

List of Contributors

Richard Broadhurst is a national expert in recreation in forests in Great Britain. He has worked for several years for the Forestry Commission as Senior Adviser on Recreation, Access and Community and he is currently Policy Officer for the Forestry Commission's National Office for Scotland.
Forestry Commission, 231 Corstorphine Road, Edinburgh EH12 7AT, UK

T.W. Mark Chambers was a postgraduate student of environmental forestry at the University of Wales during work on this project. He received an MSc for his dissertation on the subject.
School of Agricultural and Forest Sciences, University of Wales, Bangor, Gwynedd, LL57 2UW, UK

Angela Clark is an independent environmental consultant and researcher. She has specialized in the application of GIS to recreational activity and habitat studies, including a number of projects for the Forestry Commission.
Southampton Institute, East Park Terrace, Southampton SO14 0YN, UK

Professor Paul Cloke is Professor of Geography at the University of Bristol. He was previously Professor of Human Geography at the University of Wales, Lampeter, and is Founder Editor of the *Journal of Rural Studies*. His most recent of many books is *Introducing Human Geographies* (1999) edited with Philip Crang and Mark Goodwin.
School of Geographical Sciences, University of Bristol, University Road, Bristol BS8 1SS, UK

Barry Collins has been working in managerial positions in Center Parcs since 1988, first as Grounds' Maintenance Manager and since 1996 as Ecology Manager. He coordinates specialist ecological surveys and reports on the

management required for the biodiversity that is essential to the Center Parcs experience. He took a masters qualification in Countryside Management at Manchester Metropolitan University and has always been involved in agricultural and animal life in Nottinghamshire.
Center Parcs Ltd, Head Office, Kirklington Road, Eakring, Newark, Nottinghamshire NG22 0DZ, UK

Simon Evans is Head of Undergraduate Studies in Leisure, Tourism and Environment at Anglia Polytechnic University. An active researcher within the Planning Research Centre, he has published widely on issues surrounding community forestry. Research projects which he is currently involved with include sustainable tourism in Maldives and in Goa.
School of Design and Communications Systems, Anglia Polytechnic University, Victoria Road, South Chelmsford CM1 1LL, UK

Xavier Font is Tourism Lecturer at Buckinghamshire Chilterns University College (UK) and Project Officer for Tourfor, a European Union project to develop a tourism award based on environmental management systems. He has been educated in tourism marketing in Spain and the UK and delivered distance learning tourism courses in South Africa. His research interests include marketing for small tourism and recreation companies and rural tourist destinations.
Faculty of Leisure and Tourism, Buckinghamshire Chilterns University College, High Wycombe HP11 2JZ, UK

Dr C. Michael Hall is based at the Centre for Tourism, University of Otago, Dunedin, New Zealand. He is also Visiting Professor in Tourism at Sheffield Hallam University, UK and Senior Research Fellow with the New Zealand Natural Heritage Foundation, Palmerston North. He has published widely in tourism and related subjects and has a long interest in environmental history, particularly the history of the wilderness idea and the travels of John Muir.
Centre for Tourism, University of Otago, Dunedin, New Zealand

Paddy Harrop has been Environment and Communications, Policy and Support Officer for the Forestry Commission since 1997. His responsibilities include developing policy and guidance for recreation access and environment (archaeology and conservation) issues on FC land across Great Britain. He has undertaken a range of roles including recreation, planning, forest management and timber harvesting, including 3 years management of the Argyll Forest Park (Scotland).
Forestry Commission, 231 Corstorphine Road, Edinburgh EH12 7AT, UK

Dr Joan Henderson has worked for a UK tour operator and the public sector tourism administration. After this period in industry, she became a Lecturer and then Assistant Professor in the subject and joined Nanyang Technological University in 1997. She has contributed to textbooks as well as international journals and written a series of case studies on different

aspects of tourism management. Research interests include the impacts of tourism, tourism planning and tourism in South-east Asia.
Division of Marketing and Tourism Management, Nanyang Technological University, 639798, Singapore

Dr James Higham holds the position of Lecturer at the Centre for Tourism, University of Otago, Dunedin, New Zealand. His research interests in tourism include environmental management, wilderness management and the impacts of wildlife tourism. He is a founding member of the New Zealand Wilderness Research Foundation, University of Otago.
Centre for Tourism, University of Otago, Dunedin, New Zealand

Graham Hunt is currently the Director of the Forest of Mercia Project. He is an environmental planner with additional qualifications in landscape architecture and ecology. His plan-making experience with local authorities and his involvement since the early 1980s with urban and community forestry, provided a very useful background from which to approach the task of writing the type of environmental plan needed to secure community support for the proposed Forest.
Forest of Mercia, Unit 18 Rumer Hill Business Estate, Cannock, Staffordshire WS11 3ET, UK

Dr David Johnson is currently Head of Environmental Studies at Southampton Institute. He is a professional member of the Institute of Ecology and Environmental Management, the Chartered Institute of Water and Environmental Management and the Institute of Leisure and Amenity Management. Previously managing director of two consultancy and training organizations, he joined Southampton Institute in 1992. He has written extensively on the environmental impacts of leisure and tourism, and is co-editor of *Coastal Recreation Management*, published by E&FN Spon in 1996.
Southampton Institute, East Park Terrace, Southampton SO14 0YN, UK

Dr Owain Jones has an MA in Environmental Philosophy and Geography, University of the West of England; MSc, Society and Space, and PhD, University of Bristol. He is currently a Research Fellow at the University of Bristol, and Visiting Lecturer in Environmental Philosophy and Politics at the University of the West of England.
School of Geographical Sciences, University of Bristol, University Road, Bristol BS8 1SS, UK

Dr Marco Linda is a freelance consultant in the field of tourism and sustainable tourism. Sectors of specialization include marketing, planning and training (designing and coordinating training projects and lecturing). He is founder of Ecoway, a non-profit organization whose aim is to diffuse sustainable forms of tourism and ecotourism development.
Ecoway, Via delle Linfe 21.34128 Trieste, Italy

Professor Geoffrey Kearsley is Director of the Centre for Tourism and also of the Wilderness Research Foundation at Otago University. He is also presently Director of the University's Environmental Policy and Management Research Centre. Geoff is a geographer with a BA and PhD from Queen Mary College, University of London, where he began his lecturing career. He has been at Otago for the past 25 years. His current research interests are in tourism and wilderness management, urban tourism and cultural heritage with particular reference to NZ Maori.
Centre for Tourism, University of Otago, Dunedin, New Zealand

Dr Simon McArthur has 12 years experience in planning and delivering for tourism activity in highly sensitive, high profile areas throughout the world. He has worked extensively with protected area managers across Australia in tourism and visitor management, has published several books on visitor management, and has a PhD in visitor impact management. Simon's present focus is on strengthening the business performance of special interest tourism operations, and assisting tourism developments to be more imaginative in their product, and more ecologically sustainable.
Missing Link Tourism Consultants, PO Box 1461, Crows Nest, New South Wales 2065, Australia

Professor Claudio Minca is Senior Lecturer in the Faculty of Oriental Languages at the University of Venice, where he teaches Human Geography, while also teaching a course of 'Postmodern Cities and Spaces' at Venice International University. His major research interests centre around cultural issues in geography and, more especially, the meaning and role of representation in shaping tourism spaces. He is author of a book on the postmodern geography of tourism (Spazi Effimeri, 1996) and the editor of two forthcoming volumes on postmodern geography.
University of Venice, Casella Postale 179, 30100 Venice, Italy

Christopher Midgley has a degree in philosophy from McGill University in Montreal. He is an avid telemark skier, and is currently pursuing a Masters degree in Environmental Studies at York University in Toronto. His major research interests include forestry, recreation and green entrepreneurship.
Faculty of Environmental Studies, York University, Toronto, Ontario M3J 1P3, Canada

Dr Hannes Palang has been working as a researcher at the Institute of Geography, University of Tartu since 1994, obtained a MSc in environmental management from the University of Amsterdam in 1996 and a PhD in geography from the University of Tartu in 1998. His main research interests include landscape management and cultural landscapes.
Institute of Geography, University of Tartu, Vanemuise 46, 51014 Tartu, Estonia

Professor Colin Price has degrees in forestry and land use economics from Oxford University, where he also taught urban and agricultural economics.

Since 1976 he has taught environmental and forestry economics, landscape design and recreation management at the University of Wales, Bangor.
School of Agricultural and Forest Sciences, University of Wales, Bangor, Gwynedd LL57 2UW, UK

Mart Reimann has since 1996 been working as a nature conservation specialist at the Environmental Department of the Harju County Government, being responsible for management of protected areas. He has also been active as a tour leader in Põhja-Kõrvemaa. From 1999 he will be lecturing on ecotourism and recreation at the Tallinn Pedagogical University.
Environmental Department, Harju County Government, Viljandi Road 16, 11216 Tallinn, Estonia

David Russell joined the National Trust as Head of Forestry in 1986, after working with Tilhill Forestry Ltd for 10 years as a manager. He oversees the work of the Access, Coast and Countryside and Environmental Practices Advisers for the National Trust. He is currently developing guidelines on the preparation of Statements of Significance as a basis for conservation management planning across the Trust.
National Trust, 36 Queen Anne's Gate, London SW1H 9AS, UK

Dr L. Anders Sandberg teaches in the field of environmental policy and resource management. His current research interests centre on forest and environmental history in Canada. He is the editor of *Trouble in the Woods: Forest Policy and Social Conflict in Nova Scotia and New Brunswick* (1992), the co-editor (with Sverker Sörlin) of *Sustainability – The Challenge: People, Power and the Environment* (1998), and the co-author (with Peter Clancy) of *Against the Grain: Foresters and Politics in Nova Scotia* (forthcoming).
Faculty of Environmental Studies, York University, Toronto, Ontario M3J 1P3, Canada

Dr Trevor H.B. Sofield is director of the tourism programme at Murdoch University, and is a graduate in Tropical Environment Science and Geography from James Cook University. He is widely published in the area of ecologically sustainable development and tourism, and has been undertaking research into village and nature based tourism development in Nepal for the United Nations Development Programme.
School of Social Sciences, Murdoch University, Murdoch, WA 6150, Australia

Dr John Tribe is Principal Lecturer in the Faculty of Leisure and Tourism at Buckinghamshire Chilterns University College, UK. He has published books in leisure and tourism on economic and strategy issues. He is currently director of a European Union research project (Tourfor) which is investigating the use of environmental management systems to encourage sustainable recreation uses in forests with partners in Finland and Portugal.
Faculty of Leisure and Tourism, Buckinghamshire Chilterns University College, High Wycombe HP11 2JH, UK

Acknowledgements

A number of people and organizations have assisted in the production of this book in different ways. The idea for the book arose from a European conference held in High Wycombe on the theme of Environmental Management of Forest Tourism and Recreation. This in itself was part of Tourfor – a project co-funded by the European Commission (under the LIFE programme of DGXI), Buckinghamshire Chilterns University College (UK), North Karelia Polytechnic (Finland) and Estaçao Florestal Nacional (Portugal). For more information, see www.tourfor.com.

From Buckinghamshire Chilterns University College we would like to thank the Director, Professor Bryan Mogford and the Dean of the Faculty of Leisure and Tourism, Gill Fisher for their continued support and encouragement for the project. Martin Hamer, Dorette Biggs, Florin Ioras, Trevor Dixon and Nigel Griffiths were all instrumental in orchestrating the multidisciplinary aspects of the project and our research assistants Richard Vickery and Karen Yale give able support.

We would also like to thank our European partners Raimo Hulmi, Esa Etalätalo, Jussi Sommerpalo and Hanna Turunen from North Karelia Polytechnic, Finland, and Francisco Castro Rego, Maria Jóao Jesus and Elsa Teles Silva from Estaçao Florestal Nacional, Portugal.

Finally we would like to express our thanks for the help and support of Tim Hardwick from CAB *International*.

Xavier Font
John Tribe
High Wycombe

1

Recreation, Conservation and Timber Production: a Sustainable Relationship?

Xavier Font and John Tribe

Introduction

The demand for ecotourism and outdoor recreation is increasing and the pressures on land use are becoming more obvious, both in developed and developing countries. Forested land has traditionally been considered as of lower economical value than agricultural land, and the main output has been the production of timber. Yet a large part of the ecotourism experience and the recreational landscapes depends on the maintenance of forested land, and forests are crucial pockets of biodiversity conservation. Forests are part of the countryside that visitors enjoy, sometimes the purpose of the visit and other times just the setting for recreational activities, but little tourist revenue reaches forest owners, despite the fact that this revenue is much needed.

The creation of National Parks as land set aside from agriculture and forestry separated the concepts of conservation and recreation from traditional forms of economic land use (Hall, 1998). One of the main purposes of National Parks and other types of protected areas is to protect endangered and exemplar species and to encourage the natural biodiversity of the territory, and for this reason protected areas tend to be highly forested. However, protected land attracts large numbers of visitors to enjoy those natural resources, posing high pressures on the environment, and therefore only quiet enjoyment is promoted (FNNPE, 1993). Private, non-protected forests will usually be under pressure to generate the funds to make the economic operations sustainable, as well as maintain the environment and social functions. Depending on the main purpose of the forest – timber production, recreation or conservation – the forest owner will face different challenges to overcome for sustainable forest management.

Tourism and recreation will increasingly use the world's forest resources, in developed countries as buffer zones from daily urban life and in developing countries as the setting for nature tourism. Protected areas such as National Parks often find it difficult to cope with the increasing pressure of visitors, and it may be time to put in place systems whereby virgin forests outside protected areas can attract visitors. This should be done by managing their environment in a way that it pays to preserve forest and related biodiversity resources, rather than deforesting. Hence it is important to highlight examples of forest sites that have managed to combine multiple uses of forests, to consider their similarities and also their individual solutions to site-specific problems. This chapter will initially present some of the key issues faced by forest managers when bringing together forestry with tourism and recreation. It then introduces the chapters of the book organized in two sections. The first section highlights problems and challenges and the second proposes models and management solutions.

Key Issues for Multiple-use Forest Management

Tourism and recreation in forests

Forests and woodlands are part of the environment in which tourism and recreation take place. There are very few outdoor settings for recreation that do not have trees, either close up or in the background, and there are also very few tourism activities that cannot take place in a forest environment. Yet tourism and recreation in forests are not usually considered on their own but as part of outdoor recreation, because few visitors go to observe the forest itself but to carry out recreational activities in it. Listing activities that can take place will invariably be incomplete and instead activities will be reviewed against their dependency on being located in the forest.

Some of these can only take place in the forest, like watching and hunting certain birds or other animals, mushroom and berry picking, orienteering, paintballing and nature field studies, to name a few. Hunting is one of the recreation activities that generates highest revenues in non-protected forests (Sarker and Surry, 1998), yet the need for large forests and its incompatibility with other recreation make it a non-viable option for most places. Viewing or hoping to view wildlife has been ascribed one-third of the overall value of forests for recreation (Hodge, 1995), and this is a large attraction sold by tour operators as ecotourism in developing countries. Canopy walks are being introduced in some mature forests as a way of generating a special attraction within the forest that visitors are willing to pay for (Omland, 1997; see Chapters 8 and 12).

There are facilities and amenities that benefit from a forest setting because it acts as a shelter. This is especially true in hot countries where it acts as a protection against the sun. Camp sites in the Mediterranean often use woodlands as their setting because these are sources of shade, and eucalyptus plantations are sought after because there is little understorey and they repel mosquitoes. In developed countries it is recommended that accommodation outside rural settlements is screened to avoid visual impacts (see Chapter 17), and traditionally health spas in East Europe and the Alps have been located in forests because of their natural, relaxing properties. A study in Croatia found a direct relation between hotel location near forests with hotel prices and tourist demand, and the hotels interviewed stated that occupancy rates would drop in average by 20% if the surrounding forests were devastated (Horak, 1997). Highly impacting sports such as downhill skiing can be screened by forests, and sports requiring long distances, such as horse riding, cross-country skiing and running can be better enjoyed in partly forested landscapes.

A review of the suitability of forest sites for tourism will usually show that a large proportion of forested land would be adequate for tourism purposes. The Director of Forests in the German Government Forestry Office estimates that 'in about 90% of all German forests it is possible to simultaneously produce valuable timber, to protect soil, climate and watersheds, and to allow people access to the forest for recreational purposes' (Lang, 1995: 36). At present the development of forest parks for recreation depends on the importance given to the resources by the public sector and the community involvement (Skuras, 1996; Eagles and Martens, 1997; Ota, 1997; see Chapters 9, 10, 13 and 14). Yet this may change, since the majority of countries preparing guidelines for the management of tourism and recreation do so with the assumption of increased demand in the next few years (Council of Europe, 1995). In the short term this could result in an overuse of already scarce forest land for tourism and recreation, especially in tropical forests.

Besides assessing the type of forest that can accommodate tourism, the forest manager needs to understand the type of forest visitors want to go to. Research comparing public preferences for specific activities in forest landscapes has shown a clear difference between landscapes for recreation and timber production (Lee, 1990; Scrinzi *et al.*, 1995). The public believes farming and forestry help to maintain the beauty of the environment, and acknowledges the need for economic woodlands, but prefers such activities either to be located in remote areas, or to be softened somehow (Lee, 1990). Landscapes with evidence of felling or harvesting did not score high on aesthetic and functional qualities, and besides the immediate impact of felling operations, visitors do not like to see the logging waste and soil preparation. Although activities such as coppicing and selective felling have proved to be more acceptable than clearfelling, research shows the need and the difficulty in justifying human activity in woodlands. Forest and recreation managers will have to balance the needs for timber against the populistic views of woodlands which could sometimes be described as 'Bambi and Robin Hood's home'. The appeal of forests can be increased by enhancing the variety and contrasts within the area with different species contrasting in colour and form, tree age and structural diversity, making smaller clearings and thinning rather than clearfelling, varying scales of stands, creation of paths on the woodland edges, provision of recreational facilities at view-points, creation of honeypots and quiet enjoyment zones (Bostedt and Mattsson, 1995; Hodge, 1995; Jacsman, 1998).

Sustainable forest management

Forest tourism and recreation activities need to be put in the context of other uses of the forest in order to assess their complementarity or conflict. The overall goal of landowners and land users should be that of sustainable management. For the purposes of this book, sustainable development will be understood as the principle that land exploitation and preservation can be reconciled and carried out simultaneously. The classic Brundtland Commission (WCED, 1987) definition of sustainability will be used, as 'development that meets the needs of the present without compromising the ability of future generations to meet their own needs'. Sustainable Forest Management was defined at the Helsinki conference as: 'the stewardship and use of forests and forest lands in a way, and at a rate, that maintains their biodiversity, productivity, regeneration capacity, vitality and their potential to fulfil, now and in the future, relevant ecological, economic and social functions, at local, national and global levels, and that does not cause damage to other ecosystems' (The Forestry Authority, 1998: 8). This definition encompasses not only producing sustainable timber, but also catering for recreation and tourism as social and economic functions, as well as other functions shown in Fig. 1.1. Extensive research has been carried out around the specific concept of sustainable development in tourism (Mathieson and Wall, 1982; Gunn, 1994; Hunter and Green, 1995), forestry (Aplet *et al.*, 1993; Maser, 1994; Upton and Bass, 1995), and land use in

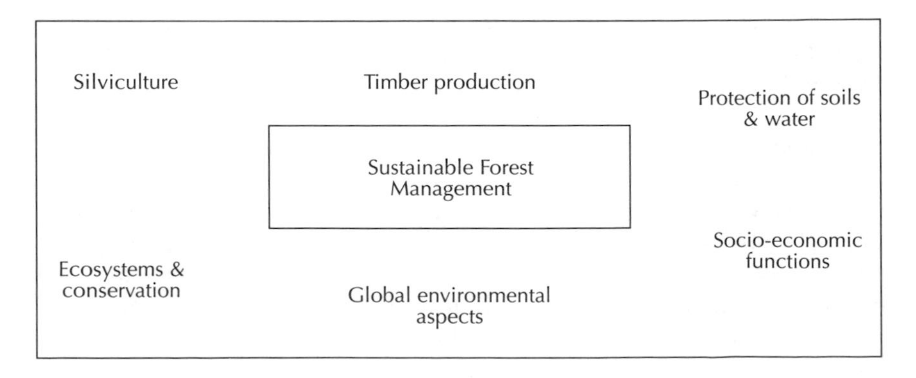

Fig. 1.1. The key aspects in forest management strategies.

general (Lele, 1991; Reid, 1995). Sustainable use of resources involves meeting economical, environmental and social requirements, which will be outlined here.

The issue of economic sustainability will include timber production and local non-timber products, as well as revenues from tourism and recreation trade, whether any of these are leased out or outsourced. Forest management issues will relate to the cycle of planting, thinning and harvesting, linked to the productivity of soils, timber quality and yield and crop rotation. In modern forestry, and especially in developed countries with high land purchasing costs, timber generates very low returns. Several countries subsidize the production of timber as a method to preserve the countryside, but forest managers are also forced to consider other sources of income (Tomkins, 1990; Dedieu, 1995; Rykowski, 1995). The importance of bringing together tourism and forestry has been recognized in forest management (JongHo *et al.*, 1997), community development (Bornemeier *et al.*, 1997) and tourism (Marcouiller, 1998; Tribe, 1998; Font, 1999). Today's forest management is much wider than simply forestry, and the non-market value of forests need to be considered.

Forest owners providing facilities for tourism and recreation need to consider the costs of building and maintaining facilities, amenities, services, infrastructure, interpretation and staffing and balance them against the potential revenues generated. Maintenance of facilities and amenities need to be considered with regards to public liability. Some landowners will decide to run the facilities themselves, others will lease out part or the whole of the operations, as is usually the case with cafés and hire shops, for example. Still other landowners will be faced with a situation where tour operators are allowed to use their land without having to pay the owner for the use, especially in areas with extensive rights of way.

Forest sites can generate revenue through entry charges (e.g. some National Parks and Nature Reserves), charges for parking, charges for participating in some activities (e.g. horse riding, archery), charges from renting equipment (e.g. skis, mountain bikes) and revenues for the provision of services (e.g. retail outlets, food and beverage, accommodation). Yet access to the outdoors is generally considered as a public right and granted by law or through tradition (Lang, 1995) at least to publicly owned land, but also sometimes to the countryside as a whole, limiting the ability to generate revenues. It is widely acknowledged that 'woodland recreation and amenity will rarely be a viable commercial venture' (Hodge, 1995: 122) and this will be usually linked to large, diverse forests that can accommodate specialist activities and can charge for entry (e.g. Chapter 17). Dupasquier (1996) found that the costs associated with recreational functions in Swiss forests varied between 15 and 200 Swiss Francs per cubic metre yield of timber, depending on whether sites simply allowed access or provided facilities for visitors, and the amount of timber production lost from dedicating land to recreation. Subsidies for setting aside agricultural land and turning it into forest are being implemented, yet this is a great cost to governments and other compensation methods are being considered (Bateman *et al.*, 1996).

The bulk of the rural tourism industry expenditure is in villages and farms. Farm tourism can provide direct revenue to the farmer by selling farm products or providing accommodation, but forest owners do not have such straightforward products to offer. The majority of research regarding the value of forests for recreation show a great difference between the value visitors allocate to the forest landscape and the income made by the owner (English and Thill, 1996; Bennet and Tranter, 1997; Dubgaard, 1998; Holmes *et al.*, 1998; Karameris, 1998; Sarker and Surry, 1998). Despite the importance of nature in general, and forests in particular as a tourism backdrop, very few hospitality businesses will own the surrounding forest or directly contribute to the upkeeping of such resources, with some reported exceptions (Caneday and Kuzmic, 1997).

The environmental sustainability of a forest relates to maintaining permanent forest cover with certain biodiversity characteristics. Some of the key issues will be preservation of endangered species, maintenance and enhancement of biological diversity (vegetation and animal), carbon sequestration, broader life support functions, environmental protection, watershed quality and climate regulation, amongst others (Manning *et al.*, 1997; Pearce, 1998). Ecological research shows that open canopy grassland is more resistant to trampling than a forb-dominated forest type (Marion and Cole, 1996), and therefore understorey environmental impacts in forests will be greater than in open land, despite the fact that the screening effect of the canopy will make it less evident. Each one of the above issues can be further qualified depending on the local forest ecology.

The social sustainability of the forest site will encompass the rights of the local community and visitors over the use of the forest. Active rights will include public rights of way, and in some instances, free recreational or social use of some parts of the forest (for example, access to religious or cultural sites that might be located in the forest). The passive rights include the right to preserve the forest for its value as a natural place, and the ability to take decisions regarding the future of that forest. For example, the right to be consulted as stakeholders in any development to ensure this is socially acceptable or does not stretch beyond the community's change tolerance level.

Forest management needs to meet many objectives, and invariably each forest will have primary objectives and secondary ones. The emphasis placed in this chapter on tourism and recreation to generate revenues is not only for the financial survival of these operations, but also because they can determine the funds available for managing environmental impacts. Yet besides the difficulty of assessing conflicts, each economic, environmental and social goal has a very different cycle. A tourist or recreational cycle usually consists of one season, at most 1 year. Timber production cycles will range from 15 to 100 years, and forest conservation cycles will span hundreds of years, if not thousands. So how can forest managers know in the short term whether they are managing their forest in a sustainable way?

International initiatives

Sustainable multi-purpose forest management is not a new concept, but the pressure placed for compliance to certain criteria is. Different initiatives are trying to deal with the issue of encouraging sustainable forest management and integrating timber production with the environmental and recreational benefits sought by society. Although these initiatives are mainly putting pressure on timber-producing forests, and not so much in conservation or recreation forests, the first are more numerous and less likely to be under government control. Also since a large amount of timber production has to be subsidized, foresters rely on meeting grant criteria for their subsistence. This section deals with some of the key international frameworks developed in the 1990s, with particular reference to the European context.

At an international level, the 1992 United Nations Conference on Environment and Development adopted the Statement of Forest Principles and the Intergovernmental Panel of Forests (IPF) was appointed. The Helsinki conference on the Protection of European Forests (1993) adopted the General Guidelines for the Sustainable Management of Forests in Europe and the General Guidelines for the Conservation of the Biodiversity of European Forests. The Dobris Assessment of Europe's Environment, published by the European Environment Agency in 1995, again drew attention to the use and management of forests and the impact this has had on the quality of the environment, and strengthened the relevance of the Helsinki declaration.

The IPF reported in 1997 to the UN, emphasizing the need to turn guidelines into National Forest Programmes with broad stakeholder input to be used as national frameworks. This is currently taking place in the form of Forest Protocols and Forest Standards (Scheiring, 1996; The Forestry Authority, 1998) and making use of the Pan-European Operational Level Guidelines devised in the Ministerial Conference on the Protection of Forests in Europe. These operational guidelines form a common framework for forest planning and practices, and consider the provision of services as a method to achieve sound economic performance, although the only direct mention to recreation and public access is made as part of maintaining 'other' socio-economic functions.

Besides the government initiatives to encourage sustainable forest management, the Forest Stewardship Council (FSC) took a market-led approach by coordinating one of the most widely known forest certification systems. The FSC goal is 'to promote environmentally responsible, socially beneficial and economically viable management of the world's forests, by establishing a worldwide standard of recognized and respected Principles of Forest Stewardship' (FSC, 1996). FSC principles and criteria emphasize the need to manage the forest for multiple purposes, using management plans that combine community, economic and environmental issues. Although tourism is not directly mentioned, it is of special importance that 'primary forests, well-developed secondary forests and sites of major environmental, social or cultural significance shall be conserved. Such areas shall not be replaced by tree plantations or other land uses' (FSC, 1996: 7), and also that 'sites of special cultural, ecological, economic or religious significance to indigenous peoples shall be clearly identified in co-operation with such peoples, and recognized and protected by forest managers' (FSC, 1996: 4).

Few policies relate specifically to tourism in forests, and current legislation relates to countryside planning in general, yet new avenues are being explored. In 1994 The Council of Europe organized the fourth Pan-European Colloquy on Tourism and the Environment, focusing on forests in Europe. Representatives from 15 countries, mostly Eastern European, signed a declaration making some recommendations on how to manage tourism and recreation in forests, which sets a milestone in the subject. Besides ratifying the general principles of sustainable management of resources, the participants recommended that 'a tourism management plan should be an integral part of the forest management plan and should be revised every ten years', and that 'institutions outside forests, but which benefit from the forest's recreational value, should participate in its tourism management costs' (Council of Europe, 1995: 62).

Different recommendations were made regarding tourism inside and outside forest protected areas. In protected areas, 'all tourist activity should be incorporated in detailed visitors' plans, which should then become an integral part of any area's management plan' (1995: 63), impacts should be regularly assessed and zoning established and all facilities should be planned and managed according to the carrying capacity of each zone, and in the majority of cases activities will be kept to the edges of protected areas. Despite the fact that more development will be allowed in forests outside protected areas, the emphasis is still on managing visitors and measuring and controlling impacts. The participants recommended that paths, barbecue sites, parking spaces, etc. should be developed as special areas designated for tourists, although larger developments such as tourist resorts and weekend houses should be situated outside forests and near already built-up areas.

The above international initiatives are the beginning of a process of implementing sustainable forest management which will carry on developing over the next few years. These form the institutional framework within which the case studies in this book have to operate. Increasingly national systems will implement methods to identify which individual forest sites accomplish the standards and will develop mechanisms to encourage those that do not, although this process will take place at different speeds across different countries.

Table 1.1. Classification of chapters.

Chapter	Author(s)	Case study	Public policy	Emphasis on impacts	Management issues	Environmental philosophy	Timber uses	Conservation uses	Recreation uses	Ecotourism	Public sector	Private sector	NGO	Values	Stakeholders/community	Conflicts	Use of models	Theoretical approaches	Practitioner views	Location
2	Henderson	X	X	X	X			X	X		X					X				Singapore
3	Reimann and Palang	X	X	X	X		X	X	X		X	X			X	X				Estonia
4	Price and Chambers	X		X	X				X		X					X	X	X		UK
5	Kearsley			X	X			X	X		X					X				New Zealand
6	Johnson and Clark	X		X	X		X	X	X		X				X	X				UK
7	Minca and Linda	X	X	X	X		X	X	X	X	X	X		X	X	X		X		Costa Rica
8	Evans	X		X				X	X	X	X				X	X				Brunei
9	Hall and Higham		X	X	X	X		X	X	X	X				X	X		X		New Zealand
10	Cloke and Jones	X			X	X		X	X					X	X	X		X		UK
11	Broadhurst and Harrop	X	X	X	X	X	X	X	X		X				X	X			X	UK
12	Sandberg and Midgley	X			X		X	X	X			X				X				Canada
13	Hunt	X			X			X	X		X	X	X		X	X	X		X	UK
14	Sofield	X	X	X	X		X	X	X	X	X	X	X	X	X	X				Nepal
15	Russell				X			X	X				X	X	X	X	X		X	UK
16	McArthur		X	X	X		X	X	X		X	X			X	X	X			Australia
17	Collins	X		X	X			X	X			X					X		X	UK

Overview of Chapters

The chapters may be classified in a variety of ways. For example there are examples from the public sector, the private sector and non-governmental organizations (NGOs). Some deal with issues of general policy and others with more site-specific management. Whilst most chapters discuss multi-benefit forests, some have an emphasis on timber production, some on recreation and some on conservation. Some authors take a theoretical approach and others are practitioners who offer practical insights. The chapters provide wide coverage in terms of geographical spread and forest types.

Table 1.1 offers some guidance to readers who have particular interests to follow. Whilst most chapters encompass both problems and possible solutions, they naturally divide into those which emphasize the problems and issues to be tackled and those which emphasize solutions. This division is reflected in the two main parts to the book – Part One: Issues and Problems and Part Two: Strategies and Solutions. Whilst the book as a whole does not purport to offer comprehensive coverage it does offer a unique insight into contemporary issues of environmental management of forest tourism and recreation across a range of contexts. Perhaps above all else the issues of community involvement and stakeholder analysis emerge as those of most significance for sustainable environmental management.

Part I: Issues and Problems

The first two chapters in Part I are case studies that illustrate the problems of competing interests and their potential effects on the future of forests in Singapore and Estonia. Joan Henderson's chapter, entitled 'The Survival of a Forest Fragment: Bukit Timah Nature Reserve, Singapore' presents an unusual case study of an area of tropical rainforest located in one of the world's most urbanized and densely populated countries. The forest environment of the reserve is typical of that which covered Singapore before its development, combining primary forest with secondary jungle. The features of special note are its diversity of tree, plant and animal life, its small size (approximately 164 ha) and its easy accessibility from all parts of Singapore. The major threats identified by Henderson to the survival of the Bukit Timah Reserve are twofold. The first is a question of literal survival. Land is in particularly short supply in Singapore and housing, infrastructure, water catchment and military needs compete strongly with recreational and conservation interests. Here Henderson warns that sites such as Bukit Timah will be sacrificed in whole or in part if it is deemed to be in the national or strategic interest. The second threat is of sustainable survival and here Henderson warns of the damaging environmental impacts caused by the heavy tourism and recreational uses placed on the site. It is in the context of these threats to the reserve that Henderson locates different stakeholder groupings, their competing interests and the roles that they need to fulfil in order to guarantee a sustainable future for Bukit Timah.

'Competing Interests on a Former Military Training Area: a Case from Estonia' is the subject of Chapter 3 by Mart Reimann and Hannes Palang. The focus of their chapter is a former military training area which since 1991 has been designated as the Põhja-Kõrvemaa Landscape Reserve. This was followed in 1997 by the designation of two special management zones and one limited management zone. Four competing interests are identified as users of the Reserve. First, nature conservation authorities who want to keep the high natural values of the area as intact as possible. Second are the former owners and their heirs for whom profitability is a key objective and most often the best way to achieve this is thought to be through logging. Third, the area has almost no human population and therefore the military regard the area as a potential site for a military training area. Finally, the area has become famous as a recreation site among the inhabitants of Tallinn.

Reimann and Palang argue that since the area has been given status as a landscape reserve, it is supposed to have protection rules as well as a management plan which enable these competing interests to be resolved. They describe how the new management plan tries to find a compromise between all these interests. The chapter explains that the prevailing opinion, shared by the nature conservation authorities as well as local communities, is that logging and military activities should be kept away from the area. However, although forestry works should be kept as limited as possible, they cannot be avoided. Similarly problems with the military are demonstrated to be delicate ones that are not easy to tackle. Reimann and Palang underline the fact that since the area belongs to a landscape reserve, nature conservation will remain the first priority. They record efforts that have been made to keep tourism, recreational and forestry activities away from the areas with the highest ecological values, alluding to the new Visitors' Centre and nature trails that encourage this. However, evidence is found to demonstrate the problems that persist in managing recreational uses. The authors show that the plan tries to give a second priority to recreation, especially so that the scientific and educational values of the area may be better realized. The chapter adds a useful perspective in its examination of the problems faced by and the methods deployed for forest management in the face of competing demands in a former Soviet-bloc country.

Chapters 4, 5 and 6 focus on impacts arising from recreation. Chapters 4 and 5 investigate the problem of crowding from the very different contexts of the UK and New Zealand. The title of the investigation by Colin Price and Mark Chambers in Chapter 4 is 'Hypotheses about Recreational Congestion: Tests in the Forest of Dean (England) and Wider Management Implications'. They note that several hypotheses purport to explain why surveys fail to detect adverse visitor response to crowding at recreation sites. Price and Chambers point to confirmation that has been found in the USA, and surveys in the Forest of Dean that provide evidence for the hypotheses. They explain how numerous interfering variables obscure relationships between crowding and satisfaction. Broadleaved woodland is found to be effective in reducing perception of crowding and displacement of crowd-averse by crowd-tolerant visitors is important and can bias results. Use of a crowd-aversion index improved satisfaction–density relationships. It is usually visitors without expectations about crowding who are insensitive to it. The results from their study in the Forest of Dean (UK) re-affirm the satisfaction–density model. Price and Chambers found strong cumulative evidence that the satisfaction of many summer visitors to the Forest of Dean was adversely affected by crowding. They conclude that forest managers should plan facilities with the adverse, though variable, effects of crowding in mind.

'Balancing Tourism and Wilderness Qualities in New Zealand's Native Forests' is the subject of Chapter 5 by Geoffrey Kearsley. The focus here is New Zealand's native forests which cover around a quarter of the country's land area. These native forests are almost entirely protected, in a more or less pristine state, by a system of National Parks and Forest Parks that emphasizes conservation above any other use, including recreation. Kearsley reviews the recreational use of these forests and their associated impacts. He finds that New Zealanders' and overseas visitors' satisfaction has been affected by a large recent increase in overseas users, although that use is presently confined, for the most part, to the more popular and easier walking tracks. Although Kearsley's study shows a high level of satisfaction with the experiences gained from back-country experiences, it also shows significant perceptions of crowding, some environmental damage and noise pollution. Kearsley notes that actual displacement, in various forms, has occurred and that there is a potential reservoir of more. The chapter concludes that traditional management methods are no longer adequate, and the consequences of largely tourist-induced crowding are beginning to impact significantly upon both traditional and recent users. Kearsley urges a change in emphasis towards a greater understanding not of the forest ecology, but of the perceptions and expectations of the users themselves. He states that the traditional wildlife focus of forest management must now be joined by a much stronger social scientific perspective, and that this is something that has not, as yet, sufficiently occurred.

Chapter 6, by David Johnson and Angela Clark, provides a 'Review of Ecology and Camping Requirements in the Ancient Woodlands of the New Forest, England'. The New Forest is the most ecologically important assemblage of lowland heath and ancient semi-natural pasture woodland in northern Europe, but its location in southern England means that visitor pressures are immense – indeed visitor numbers are in excess of 7 million per year. Camping is an important part of the recreational use of the New Forest although it has now been restricted to nine formal campsites in the forest. As the authors point out, potential impacts of camping include erosion, soil compaction, tree damage, wildlife disturbance, trampling, accidental fires, littering and vandalism. One particular site – Hollands Wood – is the focus of Johnson and Clark's case study. Their study involved an environmental appraisal of the site and comparison with similar adjacent areas of Ancient and Ornamental woodland using geographical information systems (GIS) to present the results. The appraisal listed six items of concern including impoverishment of ground flora, little natural regeneration and reductions in the amount of deadwood and woodland debris. As a result of their study, Johnson and Clark proposed a ten-point plan of design and management changes. These included the redesign of facilities, access restrictions, track resurfacing, dog bans, prohibition of kerosene-burning equipment and provision of more educational materials to encourage environmentally friendly visitor behaviour.

In their conclusions, Johnson and Clark note that moves to control access and camping in the 1960s have been instrumental in preserving the fabric of the New Forest but these have been at the expense of the 'honeypot' sites created. Environmental review and management actions can reduce unfavourable impacts at sites such as Hollands Wood. However, the long-term survival and protection of the rare habitat of Ancient and Ornamental woodland represented at the Hollands Wood site perhaps needs a more drastic solution of relocation – a solution which will require a much more difficult process of consensus building.

Chapters 7 and 8 critically examine ecotourism as a strategy for forest conservation. In each case problems associated with ecotourism surface. In Chapter 7, territoriality is the focus of Claudio Minca and Marco Linda's study – 'Ecotourism on the Edge: the Case of Corcovado National Park, Costa Rica'. They develop a theoretical approach to describe the role of tourism for local communities and their territorialities which they examine using Corcovado as a case study. Corcovado hosts one of the best-preserved tracts of Pacific Coastal Rainforest in Central America. Minca and Linda's analytical framework uses a geographical interpretation of systems theory focusing on analysis of the concepts of region and regionalization. The goal of their chapter is to identify relationships between tourist development and the other processes that have contributed to the forging of the territorialities of Corcovado and its immediate surroundings. Particular attention is paid to the relationship between tourist territorialization and pre-existing territorial activities. Analysis of these relationships enables principal areas of conflict, and of synergy, to emerge for Corcovado. It also holds the ecotourism claims of the area up for close examination.

Minca and Linda conclude that tourism development in its current evolution does not contribute to the strengthening of the territorial system of Corcovado. In particular they note that the relationship of tourist territorialization with local farmers is almost non-existent and constitutes a strong source of potential conflict. In many instances, local people have been deprived of their base source of income to preserve the natural beauty of the Osa peninsula and received very little in exchange. Minca and Linda call for a more sustainable planning approach to tourist development and offer their model of applied territoriality as a useful tool for the better planning of future development.

'Eco-tourism in Tropical Rainforests: an Environmental Management Option for Threatened Resources?' is the title of Chapter 8 by Simon Evans. Here the focus is on tropical deforestation and ecotourism as an incentive for host nations to protect rather than exploit their natural resource base – in this case tropical primary forest. Evans evaluates ecotourism as a tool of environmental management and questions the extent to which authenticity is being compromised in the long term by environmental control requirements and visitor expectations. He argues that, in the name of sustainable development, once sufficient visitation has been secured in an area to provide financial support for conservation, many additional forms of environmental control will follow. Additionally, many of these controls will be of a similar form. He sees the consequences of this as twofold. First the environmental controls may allow an intensification of visitor numbers that can be accommodated at ecotourism destinations. Second this move towards mass ecotourism will cause more adventurous travellers to seek out new, unspoilt environments to colonize. Evans warns that this cycle may eventually lead to a loss of distinctiveness and the emergence of stereotypical developments on an international scale. Developments in the

Ulu Temburong National Park in Brunei Darussalam are used to illustrate these points.

Policy is viewed as a problem rather than a solution in Chapter 9. Michael Hall and James Higham's study is 'Wilderness Management in the Forests of New Zealand: Historical Development and Contemporary Issues in Environmental Management'. Hall and Higham draw attention to the importance of forest wilderness in contemporary economic development in New Zealand. They identify a policy which they term 'New Economic Conservation'. This is aimed at maximizing the economic benefits of tourism and means that the management of forest wilderness resources has become a pressing concern given the growth in visitor markets envisaged by the 'New Economic Conservation'. Their contribution seeks to address the significance of institutional arrangements for the environmental management of tourism in the context of the forest wilderness. They use a historical methodology to outline the way in which previous sets of institutional arrangements and values have established environmental management regimes which influence subsequent policy settings and actions. They identify four main periods of institutional arrangements – the utilitarian period, the period of National Parks, the period of wilderness preservation and the current period of New Economic Conservation.

Their conclusions argue that the environmental management strategies are greatly determined by the institutional arrangements in which they are set. Present policy settings for the provision of wilderness experiences in designated areas appear to be changing in relation to demands for increased access for international and domestic visitors. However, Hall and Higham caution that while government and tourism organizations such as the New Zealand Tourist Board continue to focus on encouraging visitation, insufficient attention is being given to maintaining the resource base. Sustainable forest management practice requires that attention be given to both the demand and the supply side of tourism. By maintaining the historical focus on the visitor rather than the resource, the present economic emphasis serves to reinforce the original designation of natural areas as 'useless' land which only gains value through tourism, rather than the intrinsic value of the resource itself.

The final chapter of Part I is titled 'From Wasteland to Woodland to "Little Switzerland": Environmental and Recreational Management in Place, Culture and Time'. In this chapter Paul Cloke and Owain Jones bring social theory and environmental philosophy to bear in the analysis of the development of Camerton, a woodland in south-west England. In doing so they illustrate clearly the importance and meaning of place, culture and non-human agency in the understanding of Camerton. Their narrative description of the historical development of Camerton and its associated trees gives full voice to the complexities involved in stakeholder consultation in the determination of management and development. Indeed Cloke and Jones suggest that the trees themselves may be considered as stakeholders. It is certainly true that trees do demonstrate the key stakeholder attribute of power although to what extent they display the other key attribute of interest is more debatable. In terms of environmental management, Cloke and Jones enable us to tune into the rich complexity of stakeholder interest. This is in stark contrast to the glib way in which stakeholder interest and consultation can sometimes be reduced to one or two public meetings with a view to consensus. Cloke and Jones underline the problem of contestability and the rich reality which defies consensus. Whilst it does not offer any easy solutions to environmental management their chapter does emphasize the difficulties of achieving true community and stakeholder consensus on management and development.

Part II: Strategies and Solutions

The chapters in Part II all emphasize the role of community and stakeholders whilst presenting different models and examples of environmental management. In Chapter 11, Richard Broadhurst and Paddy Harrop provide us with insights into the way in which forest tourism is developed in Great Britain by the Forestry Commission that make an interesting contrast to the issues raised by Hall and Higham in Chapter 9. 'Forest Tourism: Putting Policy into Practice in the Forestry Commission' initially investigates the background issues to forest tourism. It reviews studies which attempt to quantify the value of forest tourism whilst pointing out the dangers of taking a narrow economic view of things. Their brief history of British forests emphasizes the very different factors that drove policy in 1919 compared with those that had emerged 80 years later in 1999. Then, the creation of strategic domestic timber supplies was paramount. Now, multi-benefit forestry gives due weight to tourism and recreation activities. The establishment of the first National Forest Park in Scotland in 1935 is seen by Broadhurst and Harrop as a crucial event marking this change in perception and emphasis.

The chapter explains policies and process that guide forest management that is overseen by the Forestry Commission, and demonstrates through the case study of Glen Affric (Scotland) how policies are put into practice in relation to tourism and recreation, emphasizing the important part played by the forest design plan. It concludes by considering the key factors that influence policy that is under continual development by the Forestry Commission and the importance of consultation is stressed. Policy is underpinned by developments at the national level (UK Forestry Standard), European Level (Pan-European Criteria) and international level (Forest Stewardship Council Principles), each of which gives consideration to tourism and recreation. The chapter ends by emphasizing the increasing role that tourism has to play noting that 'through forest tourism we have the opportunity to join up our thinking and practice, so that tourism increases enjoyment as well as understanding of our relationships with trees, woods and forests, and the essential role they play in sustaining life on our planet'.

The problem at the heart of L. Anders Sandberg and Chris Midgley's study in Chapter 12 is one of dualism. They identify this as the typical North American pattern of spatially separated uses of forests. They observe that forest recreation in the province of Ontario, Canada, is typically focused in parks and preserved areas with very little integration of such activities with other economic pursuits, such as forest harvesting. They note that where integration occurs, it is usually as a result of logging in designated park areas, which is generally frowned upon by the public, abhorred by environmentalists and carefully shielded from the visiting public by forest managers. Sandberg and Midgley explain that for the most part forest harvesting occurs on company-leased provincial Crown forest lands. On such lands recreational activities are not promoted generally, but may occur incidentally, such as is the case with fishing, hunting and snowmobiling.

In their chapter 'Recreation, Forestry and Environmental Management: The Haliburton Wildlife Reserve, Ontario, Canada', they tell the story of private forest owner Peter Schleifenbaum who is trying to counter this trend. The owner of the 21,751 hectare Haliburton Forest and Wildlife Reserve in Ontario is attempting to build an environmental management strategy that combines recreational developments with the rehabilitation of a degraded forest and the cultivation of trees for value-added manufacture. His strategy is built on the effective use of the rights that go with private ownership, and a combination of other political, economic, public relations and market measures. Sandberg and Midgley describe these strategies and the operations of the Reserve and assess their advantages, drawbacks and lessons for the combined use of recreation and forestry more generally.

Chapters 13 and 14 each respond to some of the issues of community participation and guiding values raised in Part One. In Chapter 13, Graham Hunt explains the process of 'Writing an Environmental Plan for the Community Forest of Mercia, England'. The Forest of Mercia is one of 12 new forests being created as part of a nation-wide programme of Community Forests. This programme is an important environmental initiative in England. The community forest programme seeks to utilize multi-purpose forestry as a means of bringing about significant landscape enhancement on the edge of major towns and cities in England. In turn these improvements will create the opportunity for the delivery of a wide range of social, economic and environmental benefits. A prerequisite of all of these plans is that they must be based upon widespread public support and voluntary participation by landowners. The plans themselves will have no statutory basis and therefore will be heavily dependent on linkages to existing plans and policies. All of these presumptions have direct implications for the type of plan-making process that has had to be adopted to produce acceptable plans. Hunt's examination of the Forest of Mercia Plan shows how an innovative and visionary environmental plan can be developed with widespread community support and be closely integrated with the social and economic agendas within the plan area.

Chapter 14 turns attention to the forests of Nepal which are an important resource for tourism and recreation and provide essential materials and products for the survival of its rural communities. 'Forest Tourism and Recreation in Nepal' is the title of Trevor Sofield's chapter where it is noted that a forest protection regime has existed for centuries in royal forest reserves, sacred sites' forests and community forests. In terms of contemporary practice, Sofield describes a growing alliance between protected areas, forests and tourism in Nepal. In particular he notes that 14% of the country has protected area status with the twin objectives of safeguarding biodiversity and maximizing tourism benefits. In examining Nepal's approach to the conservation and utilization of its forest resources Sofield challenges two popular assumptions. First he reviews the 'deforestation crisis' and concludes that the crisis was overstated and that responses provoked have led to a significant mitigation of the problem. Second he examines 'the tragedy of the commons' noting that anthropological literature vigorously refutes the assumption that local people are ignorant of good forestry practice.

The chapter then describes two institutionalized examples of good practice using sustainable approaches to the environmental management of forests for tourism and recreation. These are the Annapurna Conservation Area Project and the Parks and People Programme. According to Sofield, both of these projects incorporate strong elements of community participation in forestry management and both therefore have the potential to encourage sustainability in economic, ecological and socio-cultural contexts. He concludes that the role of tourism in underpinning these efforts by providing a sound economic foundation which can penetrate local communities is essential to the sustainability of conservation and protection of the forests and biodiversity of Nepal.

In Chapter 15, 'Planning for the Compatibility of Recreation and Forestry: Recent Developments in Woodland Management Planning within the National Trust', David Russell, writing from a practitioner perspective as Head of Forestry for the National Trust (England and Wales), notes that within the last 20 years UK forestry has become a multi-purpose enterprise. Providing for recreation has emerged as a major component of forest management supported by substantial grants of public money. Russell is confident that woods and forests can absorb many visitors and are resilient to many forms of recreational activity.

In Russell's view, woods are much more than an arena for leisure time activity. Woods and trees have special qualities which stir profound emotional responses and understanding of this is important. In response to this, management planning advocated by Russell is based on a 'Statement of Significance' which summarizes all the ways in which the wood is used and valued. This statement is seen as an essential basis for a continuing dialogue with stakeholders and for building compatibility between all the components in a genuinely multi-purpose forestry enterprise.

Chapters 16 and 17 provide two models for comprehensive environmental management. Simon McArthur's chapter, 'Beyond Carrying Capacity: Introducing a Model to Monitor and Manage Visitor Activity in Forests', reviews conventional approaches to managing visitors including Carrying Capacity, the Visitor Impact Management model, the Visitor Experience and Resource Protection model, the Visitor Activity Management Programme and the Limits of Acceptable Change (LAC) model. McArthur argues that all of these models suffer from a failure to establish sufficient stakeholder support largely because the culture inherent in the models is not attuned to attracting wider stakeholder involvement. In response to these criticisms a new model TOMM (Tourist Optimisation Management Model) is described. It is explained that most of the components of TOMM are similar to the LAC, but that while LAC is strongly focused on the decision-making process, TOMM has more emphasis on the contextual analysis and monitoring programme. Additionally TOMM is designed to serve a multitude of stakeholders with a multitude of interests, and can operate at a regional level over a multitude of public and private land tenures.

McArthur concludes that the philosophy of TOMM is particularly valuable for use in forests where values are diverse and thus competition for different outcomes is intense. Reviewing the three examples where the TOMM has already been implemented (Kangaroo Island, Dryandra Woodland and Banff National Park) he suggests that there is real merit in not only using the TOMM to manage visitors in forests, but in integrating the essence of TOMM into broader environmental management of forests. He sees a place for TOMM as a general model for monitoring and managing forests. He notes that forest managers across Australia have already jointly identified indicators and targets for the conservation and sustainable management of temperate and boreal forests, as part of the 'Montreal Process'. While this initiative and TOMM are, according to McArthur similar in their first two stages (context analysis and monitoring), TOMM is recommended for its simple performance standard (acceptable range) and simple reporting system that makes it accessible to a wider range of stakeholders.

Finally, Chapter 17 focuses on the uses for forest tourism and recreation of Environmental Management Systems (EMS). In the chapter titled 'Implementing Environmental Management Systems in Forest Tourism: the Case of Center Parcs', Barry Collins explains that Center Parcs is implementing this standard across all 13 of its villages, in all five European operating countries, and gives examples of the results in the UK. This chapter offers a brief overview of the benefits of a complete management system for landscape management and biodiversity. The International Standard for Environmental Management Systems (ISO 14001) is used as an example to focus predominantly on the organization's management system, showing compliance to environmental legislation and a documented management system that completes the cycle of a plan, do, check and act. The chapter outlines the corporate philosophy of Center Parcs and records the tangible outcomes achieved by Center Parcs with regard to wildlife conservation as a result of implementing a complete management system. Center Parcs shows how using an EMS helps to achieve continuous environmental improvement, and Collins refers to the many accolades that have resulted from the efforts of Center Parcs and concludes that the company is now recognized as a sector champion for biodiversity in the leisure industry.

Conclusions

The chapters in this book outline a variety of situations and experiences for the forest owner and recreation manager to draw from, with some common principles relating to sustainable tourism and recreation management in forests.

Site owners are responsible for the management of their resources, their environmental quality and visitor safety while on site. Traditionally land managers have considered that the way to minimize the impacts of tourism or recreation in their land is to not encourage visitors. Current pressures on the countryside have forced landowners to acknowledge that visitors will not disappear by just ignoring them, and that it is more beneficial to provide managed areas in order to concentrate usage. The majority of visitors are happy to stay in trails and logging roads and recreational areas catering for them. Environmental management techniques and overall systems have been devised for this purpose, bringing together countryside, forestry and recreational needs (Liddle, 1997; Hammitt and Cole, 1998; Tribe *et al.*, 2000).

The overall sustainability of the tourism and recreation operations should be assessed at four levels: the visitor impact on the immediate tourism site, the off-site impacts of running the site, the contribution to improvement of non-tourism environmental goals (sustainable forest management) and the contribution of site activities to the sustainability of the local economy. Most operations concentrate only on the first heading, yet truly sustainable tourism and recreation management in the forest should address all four. The contributions to this book show that stakeholder and community involvement in responsible forest management are key to the sustainability of the resources.

No definitive rules can be laid out relating to what is a negative impact of tourism in forests, and in general this will not depend on the impact itself, but the context in which it happens. Cattle grazing of forests is considered part of the causes of deforestation in Poland, but the New Forest (England) pasture woodlands are heritage landscapes in themselves. Berry and mushroom picking are considered as positive in Finland, accepted in Spain and negative in the UK. Slash and burn farming is part of the Finnish heritage being recovered, but one of the key reasons for deforestation in South-east Asia. Forest fires are the single largest potential impact of recreation in Mediterranean and Australian forests, yet considered part of the natural cycle in Scandinavia.

All the above suggest that international initiatives to encourage sustainable forest management need to be locally interpreted not only by national bodies, but also by the forest site manager (Upton and Bass, 1995). Site specific environmental management systems, devised by the forest manager with stakeholder involvement are powerful tools to be both sensitive to the local conditions and to pick up internationally agreed environmental priorities (Tribe *et al.*, 2000).

References

Aplet, G.H., Johnson, N., Olson, J.T. and Sample, V.A. (ed.) (1993) *Defining Sustainable Forestry.* Island Press, Washington.

Bateman, I.J., Diamand, E., Langford, I.H. and Jones, A. (1996) Household willingness to pay and farmers' willingness to accept compensation for establishing a recreational woodland. *Journal of Environmental Planning and Management* 39, 21–43.

Bennet, R. and Tranter, R. (1997) Assessing the benefits of public access to the countryside. *Planning Practice and Research* 12, 213–222.

Bornemeier, J., Victor, M. and Durst, P.B. (eds) (1997) Ecotourism for forest conservation and community development. Proceedings of an international seminar held in Chiang Mai, Thailand, 28–31 January 1997, RECOFTC report, No. 15, Regional Community Forestry Training Center for Asia-Pacific, Bangkok.

Bostedt, G. and Mattsson, L. (1995) The value of forests for tourism in Sweden. *Annals of Tourism Research* 22, 671–680.

Caneday, L. and Kuzmic, T. (1997) Managing the diverse interests of stakeholders. *Parks and Recreation* 32, 118–126.

Council of Europe (1995) *Forests in Europe: Proceedings from the 4th Pan-European Colloquy on Tourism and the Environment,* Warsaw, 20–21 September 1994. Council of Europe Press, Strasbourg.

Dedieu, J. (1995) The functions of Europe's forests – public expectations, the dangers ahead and the economics of management. In: *Forests in Europe: Proceedings from the 4th Pan-European Colloquy on Tourism and the Environment,* Warsaw, 20–21 September 1994. Council of Europe Press, Strasbourg, pp. 17–18.

Dubgaard, A. (1998) Economic valuation of recreational benefits from Danish forests. In: Dabbert, S., Dubgaard, A., Slangen, L. and Whitby, M. (eds) *The Economics of Landscape and Wildlife Conservation.* CAB International, Wallingford, pp. 53–64.

Dupasquier, P. (1996) Cout de la fonction de delassement en forêt. *Schweizerische Zeischrift fur Forstwesen* 147, 572–583.

Eagles, P. and Martens, J. (1997) Wilderness tourism and forestry: the possible dream in Algonquin provincial park. *Journal of Applied Recreation Research,* 22, 79–97.

English, D.B. and Thill, J.C. (1996) *Assessing regional economic impacts of recreation travel from limited survey data.* Research note SRS-2- Southern Research Station, USDA Forest Service, Asheville, USA, Southern Forest Experiment Station, USDA Forest Service.

FNNPE (1993) *Loving them to death?: Sustainable Tourism in Europe's Nature and National Parks.* Federation of Nature and National Parks of Europe, Grafenau.

Font, X. (1999) Tourism, recreation and the value of British woodland. *Symposium on British Tourism: The Geographical Research Frontier*, 21–23 September. RSG/IBG, Exeter.

Forest Stewardship Council (1996) *Principles and Criteria for Forest Stewardship*, Revised March 1996, edited October 1996. Forest Stewardship Council, Oxaca.

Gunn, C.A. (1994) *Tourism Planning: Basics, Concepts, Cases.* Taylor and Francis, New York.

Hall, C.M. (1998) Historical antecedents of sustainable development and ecotourism – new labels in old bottles? In: Hall, C.M. and Lew, A.A. (eds) *Sustainable Tourism: A Geographical Perspective.* Longman, Harlow, pp. 13–24.

Hammitt, W.E. and Cole, D.N. (1998) *Wildland Recreation: Ecology and Management.* John Wiley & Sons, New York.

Hodge, S. (1995) *Creating and Managing Woodlands Around Towns.* Forestry Commission Handbook 11, HMSO, London.

Holmes, T., Alger, K., Zinkhan, C. and Mercer, E. (1998) The effect of response time on conjoint analysis estimates of rainforest protection values. *Journal of Forest Economics*, 4, 7–28.

Horak, S. (1997) Influence of forested area in hotel vicinity on hotel accommodation prices. *Turizam* 45, 125–138.

Hunter, C. and Green, H. (1995) *Tourism and the Environment: A Sustainable Relationship?* Routledge, London.

Jacsman, J. (1998) Konsequenzen der intensiven Erholungsnutzung fur die Walder im stadtischen Raum. *Schweizerische Zeitschrift fur Forstwesen* 149, 423–439.

JongHo, K., KyuHun, K. and ChinKyu, L. (1997) A study on the establishment of classification index for choosing the objective areas of mountain village region development project. *Journal of Forest Science* 55, 105–124.

Karameris, A. (1998) Abschatzung der Erholungsbelastung verschiedener Waldkomplexe mit Hilfe theoretischer Modelle. *Schweizerische Zeitschrift fur Forstwesen* 149, 105–120.

Lang, W. (1995) Experiences in the recreational management of forests with various functions. In: *Forests in Europe: Proceedings from the 4th Pan-European Colloquy on Tourism and the Environment*, Warsaw, 20–21 September 1994. Council of Europe Press, Strasbourg, pp. 35–37.

Lee, T. (1990) What kind of woodland and forest do people prefer? In: Talbot, H. (ed.) *People, Trees and Woods. Proceedings of the 1989 Countryside Recreation Conference, Heriot-Watt University, Edinburgh, 19–21 September.* Countryside Recreation Advisory Group, Bristol, pp. 37–52.

Lele, S.M. (1991) Sustainable development: a critical review. *World Development* 19, 607–621.

Liddle, M. (1997) *Recreation Ecology.* Chapman & Hall, London.

Manning, R.E., Valliere, W. and Minteer, B. (1997) Environmental values, environmental ethics and national forest management: an empirical study, General Technical Report N. NE-232. Northeastern Forest Experiment Station, USDA Forest Service, Radnor, pp. 216–222.

Marcouiller, D.W. (1998) Environmental resources as latent primary factors of production in tourism: the case of forest-based commercial recreation. *Tourism Economics* 4, 131–145.

Marion, J.L. and Cole, D.N. (1996) Spatial and temporal variation in soil and vegetation impacts on campsites. *Ecological Applications* 6, 520–530.

Maser, C. (1994) *Sustainable Forestry: Philosophy, Science and Economics.* St. Lucie Press, Delray Beach.

Mathieson, A. and Wall, G. (1982) *Tourism: Economic, Physical and Social Impacts.* Longman, Harlow.

Omland, M. (1997) Exploring Ghana's tree tops. *People and the Planet* 6, 28–29.

Ota, I. (1997) Comparative study of national forest recreational activities between Japan and the United States. *Natural Resource Economic Review, Kyoto University*, No. 3, pp. 29–58.

Pearce, D. (1998) Can non-market values save the tropical forests? In: Goldsmith, F.B. (ed.) *Tropical Rain Forest: A Wider Perspective.* Conservation Biology Series, Chapman & Hall, London, pp. 255–267.

Reid, D. (1995) *Sustainable Development.* Earthscan, London.

Rykowski, K. (1995) Recreation in forests as part of the multipurpose forestry concept. In: *Forests in Europe: Proceedings from the 4th Pan-European Colloquy on Tourism and the Environment*, Warsaw, 20–21 September. Council of Europe Press, Strasbourg, pp. 11–12.

Sarker, R. and Surry, Y. (1998) Economic value of big game hunting: the case of moose hunting in Ontario. *Journal of Forest Economics* 4, 29–60.

Scheiring, H. (1996) Eine funktionenorientierte, integrale Waldwirtschaft. *Forstwissenschaftliches Centralblatt* 115, 206–212.

Scrinzi, G., Floris, A., Flamminj, T. and Agatea, P. (1995) *Un Modello di Stima Della Qualita Estetico-fuzionale del Bosco.* ISAFA Comunicazioni di Ricerca dell'Istituto Sperimental per l'Assestamento Forestale e per l'Alpicoltura, ISAFA, Trento.

Skuras, D. (1996) Regional development of forest recreation facilities: planning decisions in Greece. *Scottish Agricultural Economics Review* No. 9, 57–66.

The Forestry Authority (1998) *The UK Forestry Standard: The Government's Approach to Sustainable Forestry.* Forestry Commission, Edinburgh.

Tomkins, J. (1990) Recreation and the Forestry Commission: the case for multiple-use resource management within public forestry in the UK. *Journal of Environmental Management* 30, 79–88.

Tribe, J. (1998) Tourfor: an environmental management systems approach to tourism and recreation in forest areas. In: Hall, D. and O'Hanlon, L. (eds) *Rural Tourism Management: Sustainable Options.* International Conference Proceedings, Scottish Agricultural College, pp. 561–577.

Tribe, J., Font, X., Griffiths, N., Vickery, R. and Yale, K. (2000) *Environmental Management of Rural Tourism and Recreation.* Cassell, London.

Upton, C. and Bass, S. (1995) *The Forest Certification Handbook.* Earthscan, London.

World Commission on Environment and Development (WCED) (1987) *Our Common Future.* United Nations, New York.

The Survival of a Forest Fragment: Bukit Timah Nature Reserve, Singapore

2

Joan C. Henderson

Introduction

This chapter considers the case of Bukit Timah Nature Reserve in Singapore, an area of approximately 164 ha of forest and the last remaining tract of rainforest left in the country. The Nature Reserve is a relatively small forest and has experienced some disturbance, but it has survived in one of the most densely populated countries in the world and retains much of Singapore's original biodiversity. Nevertheless, there is some uncertainty about its future as the population of Singapore increases and the shortage of land becomes even more acute whilst heavy use of the forest by local residents and tourists is causing some damage.

Firstly, the history of the site and its particular characteristics are summarized, followed by a discussion of current concerns. The impacts of increasing recreational use, both positive and negative, are examined and the current system of management is outlined. Existing strategies are described and those proposed for the future examined. Prospects are assessed within the context of Singapore's planning and development policy for the whole island which reflects a concern about land shortage and making the most efficient use of scarce resources. Scenarios based upon the perspective of the different stakeholders are then examined, and a series of actual and potential conflicts are highlighted. Securing the cooperation and commitment of all interested parties is seen as essential to the forest's survival, alongside the continuation of existing good practice and introduction of new measures where appropriate.

Bukit Timah represents a small forest reserve within an urban and Asian context and its study provides some interesting insights into the challenges of managing such a site. It should be noted, however, that Bukit

Timah is not typical of Asian tropical rainforests in general on account of its size and use, being unaffected by the problems of commercial logging, palm oil plantation, transmigration plans and government resettlement schemes for indigenous populations characteristic of its neighbours such as Indonesia (Edwards *et al.*, 1996). Nevertheless, like many of these, the forest is affected by steadily increasing visitors – although these tend to be domestic residents rather than overseas tourists. Reaching a balance between the needs of the forest and its visitors remains a common regional problem, but other issues of forestry management for sustainable development such as those discussed by Sharma (1992) and D'Silva and Appanah (1993) rarely apply.

A Historical Perspective

As the highest point on the island of Singapore at 519 feet, Bukit Timah was a recognized landmark to European settlers of the 19th century. Yet it was rarely visited by them and considered an area of inhospitable terrain inhabited by a band of gambier farmers over whom it was difficult to exercise any control. These circumstances changed as the road network was extended and accessibility improved so that by the middle of the century it was possible to reach the summit fairly easily with the installation of facilities such as seats and tables and the construction of an official bungalow for private rental.

Bukit Timah became a popular site for leisure visitors, but also was of great interest to scientists whose studies of the geology, flora and fauna of the area are well documented. At the same time, it was exploited by the timber trade whose activities caused large-scale forest clearance throughout the island, reports suggesting that the percentage of forest land dropped from 60% in 1848 to 7% in 1882. Hence, there was 'early clearance, or at least man-made disturbance, on the Hill, so it is doubtful if more than a small part of the Bukit Timah forest can be called "virgin" or primary' (Lum and Sharp, 1996: 17).

Awareness of the resulting environmental damage led to calls for protection and Bukit Timah Hill became one of the first official forest reserves in the late 19th century, despite the fact that only about 30% of the 343 ha could be described as forested, under the control of a newly established Forest Department with responsibility later passing to the Collector of Land Revenue at the Land Office. The early 20th century saw further demands made on the land and its resources with granite quarrying, road and railway expansion and military activity as well as dairy farming. Boundaries were revised in the 1930s and the reserve area shrank to about 80 ha over which the Botanic Gardens Director exercised authority. Innovations around this time included the introduction of an extended path network, the cataloguing and labelling of trees, the installation of maps at the eight shelters and the building of an artificial pool.

After the Second World War, when Bukit Timah Hill was the location of a fierce battle between the colonial powers and the invading Japanese forces for Singapore, supporters lobbied for action to conserve what was left of the forest against inroads from development and granite quarrying in particular. The Nature Reserves Ordinance of 1951 formally recognized a number of reserves as important to the preservation and study of Singapore's native flora and fauna with the appointment of a Board of Trustees to supervise their administration. An area of just over 1600 ha to the east was designated the Central Water Catchment Forest, alongside the Bukit Timah Reserve. The Board faced a challenging period, especially after independence and the foundation of the Republic of Singapore in 1965 when the country experienced rapid change as a consequence of its pursuit of economic development and modernization which led to the disappearance of much open space.

Such spaces continued to be built on in the later decades of the 20th century with Bukit Timah facing encroachment along much of its border. A particular blow took place in 1986 when a six-lane motorway (the Bukit Timah Expressway or BKE) was constructed through the central forest area, effectively separating Bukit Timah from its Central Water Catchment Forest hinterland and creating a reserve that was bounded by roads, a railway, housing development and a rifle range. There were also internal pressures as more people were attracted to Bukit Timah, numbers rising from 78,000 in 1987 to 140,000 by 1995 (Lum and Sharp, 1996). It seems likely that these were mainly residents, most of whom lived in high-rise accommodation blocks, seeking an escape from the urbanized environment of modern Singapore. Doubts began to be expressed about whether the forest could survive and there was a growing realization, both amongst the public and officials, of the importance of it doing so.

The Parks and Recreation Department and Nature Reserves Board recognized the threat to the forest's health and future and began to consider new strategies. The government too responded by establishing a National Parks Board in 1990 to promote, develop and manage the reserves and other parks. The Board was also to help preserve the nature reserves as 'a sanctuary for wildlife, a place for plant and animal conservation, as well as a resource for education and outdoor recreational activities' (National Parks Board, 1999). In 1996, The Parks and Recreation Department of the Ministry of National Development was incorporated into the National Parks Board which is currently the trustee for the Bukit Timah Nature Reserve, now enlarged to 164 ha (400 acres).

The Singapore government also became more interested in and committed to conservation matters. The Green Plan was published in 1992 whereby 5% of the island was to be dedicated to nature conservation while Action Programmes presented in 1993 listed a total of 19 sites for protection, the Bukit Timah and Central Water Catchment Nature Reserve being the largest.

Bukit Timah has thus undergone several changes in its size and administration over the years since Singapore was claimed for the British by

Stamford Raffles in 1819. It has also been subject to adverse impacts and the demands of competing and conflicting land uses which have disturbed the environment and threatened its survival as a nature reserve. The special features of the forest, some of the impacts experienced there and the most recent plans for its conservation and management are now considered.

Characteristics of the Forest

Bukit Timah's distinguishing characteristics are its small scale, close proximity to the city centre which is only 12 km away and great diversity of plant and animal life as well as popularity. According to Corlett (1995), 'it is visited daily by hundreds of people – walkers, joggers, nature lovers, school groups and tourists. It provides a research area for many Singapore-based and visiting scientists, and has featured in numerous scientific publications: Bukit Timah may well be the best studied forest area in Southeast Asia. It is also probably the oldest rain forest reserve in the region, if not the world' (p. 2). As such, it has an important role to play internationally as an 'experiment – unplanned and incompletely documented – on the effects of fragmentation and isolation on a species-rich rain forest biota' (p. 3).

For the visitor with a more general interest, the attraction is the opportunity to experience the natural environment as it used to be in Singapore. In the words of the current visitor guide (National Parks Board, undated), 'Paths weave through small granite hills covered in dense tropical foliage, rising upward to offer a magnificent view at journey's end. Along footpaths, visitors may see an astonishing variety of plant, animal and insect life, typical of a humid equatorial climate. By virtue of its location on the equatorial belt, the Nature Reserve has one of the richest, most diverse ecological systems in the world'.

Despite some disturbance, there has been no major clearance and forest cover generally remains intact to create a lowland evergreen landscape typical of the Malay Peninsula, combining primary forest with secondary jungle (Singapore Science Centre, 1995). *Dipterocarpacea* predominate and whilst reaching heights of 50 m, these trees are vulnerable to high winds and tropical storms because of their shallow root systems and the steep topography. Several fall each year which encourages the development of a series of secondary species such as *Macaranga*. Understorey plants adapted to the low light intensity grow on the forest floor with rattans and other climbers found on the edge of clearings. Epiphytic ferns and lianes are another feature with dense carpets of ferns occurring in some of the more extensive clearings.

Under the canopy, the forest also supports many of the smaller animals native to Singapore including the long-tailed macaque, plantain and slender squirrels, common treeshrew, flying lemur, anteater and mousedeer. Rare and scarce birds are also present like the lesser cuckoo-shrike and the blue-crowned hanging parrot, in addition to other species during the migratory season. There are numerous varieties of lizards, snakes, insects and amphibians found on land and in the water. Larger animals are now extinct, however, and the BKE has separated others from their water supply (Chua, 1993).

In terms of facilities, visitors are provided with four trails, walking times ranging from 20 to 80 min, and a cycling route which takes between 30 and 60 min to complete. There is a metalled road leading up to the summit where telecommunications towers are located, and eight huts and a Visitor Centre in the Reserve which is easily accessible by public transport or private car. Boundaries are formed by the BKE in the east, the Malayan Railway in the west, the Bukit Timah Rifle Range in the south and residential development in the north. The Bukit Timah Shopping Centre is only a 10-minute walk from the entrance to the reserve.

Bukit Timah thus displays several unique features. It offers great diversity and complexity of species and, together with the Central Water Catchment area, is home to over 50% of Singapore's remaining native flora and supports fauna of much interest and variety. Its location also marks it out as distinctive, allowing it to be described as an urban forest in a country characterized by its high population density, degree of urbanization and extensive development.

Current Concerns

The nature of the site and its condition have attracted the attention of conservationists since it was first identified as an area of importance, but the construction of the BKE increased awareness of its vulnerability and demands for protection. The motorway left Bukit Timah as an island, divided from its hinterland of the Central Water Catchment Forest and the source of seeds and animal breeding partners to replace those in danger of extinction. There was a fear that this would contribute further to the degradation of primary forest and intensify secondary growth. Briffet (1990, pp. 47–48) describes the actual and possible disturbances as follows:

> **1.** The construction of the Bukit Timah Expressway has cut off this forest from the larger water catchment forest area, reducing the migratory interflow of the flora and fauna.
> **2.** The general drying up of the forest threatens some of the rare and scarce palms, ferns and freshwater life which is now in danger of extinction.
> **3.** The heavy quarrying has resulted in many landslides over the years causing several streams to be diverted or disrupted, and endangering rare freshwater life.
> **4.** During the last few years, many people bring their unleashed dogs for walks up the hill and this poses problems and disturbances for ground-dwelling animals.
> **5.** Despite the presence of wardens, poaching still goes on and several of our

members have seen bird catchers and animal trappers operating in the area. Certain peripheral areas are being used as illegal rubbish dumps.
6. Continued usage of the reserve by large numbers of unsupervised and unguided hikers causes erosion and litter problems.
7. The present tarmac road is constantly used by police, military, telecoms and contractors vehicles which create disturbance problems.

These problems persist and some have intensified in the past decade, especially those related to the impact of visitors who now arrive in very large numbers. These visitors cause erosion of footpaths and soil compaction when wandering off the marked trails, as well as disturbance to plant and animal life, and it could be argued that the Reserve's carrying capacity is being exceeded.

There is a considerable literature on tourism carrying capacity, defined by Mathieson and Wall (1982) as 'the maximum number of people who can use a site without an unacceptable alteration in the physical environment and without an unacceptable decline in the quality of the experience gained by visitors' (p. 21). Carrying capacities can be expressed in physical, psychological (or perceptual), social and economic terms (Hunter and Green, 1995) and this classification suggests the complexity of the concept and practical difficulties of its application. Problems of quantification and measurement arise with much depending upon the nature of tourists, the site and any resident population. Capacities might also change over time as alterations take place in determinants such as accessibility, and there is the additional problem of who decides when enough is enough.

Despite these weaknesses, carrying capacity acts as a reminder that a destination has limits and exceeding these will threaten its well-being and future existence. Bukit Timah clearly is constrained by its size and lack of space for expansion, and the forest environment and ecosystem is susceptible to damage by human presence. Socio-psychological factors also mean that tolerance will be tested and the visitor experience devalued if too many people are present. Economic considerations are less applicable as the forest is not a generator of income.

Without any detailed statistics, the discussion of carrying capacity cannot be pursued further; however, it remains an important issue to be addressed given the growing environmental movement in Singapore, as evidenced by membership of organizations like The Nature Society. Such trends might be seen as favourable in providing support for continued conservation, but they will also lead to the Reserve becoming even more popular and pose serious challenges for those in charge of its management. Carrying capacity theory should inform planning and management decisions, with its message that there are limits to growth and these must be recognized in the interests of sustainability. These questions are returned to at a later stage in the chapter.

An additional major concern is the growth of residential property on the Reserve's boundaries, discussed in a *Straits Times* article (1996) aptly entitled

'Nature's New Neighbours – Forests on One Side, Condos on the Other'. An architecture lecturer and conservationist is reported as claiming that such development has created more degradation in the last 6 months than in the past 10 years as 20,000 new residential units have been built on Bukit Timah's perimeter.

Impacts of Recreation Use

The Reserve is thus subject to a series of pressures from outside forces and internally as a result of increased visitor numbers. Recreational activity often impedes conservation and human presence in the forest may introduce light and air which disturbs the native species requiring shade and humidity. Erosion of heavily used paths is a serious problem, as already noted, and the footpath network exacerbates the 'edge effect' by exposing more of the forest to the external environment while some walkers stray off the marked routes. Visitors bring litter and may be tempted to feed the numerous monkeys, and the presence of dogs can also cause damage. These impacts do vary in intensity, however, and are more acute during the peak periods of weekends and public holidays; on a mid-week day, the Reserve is usually relatively quiet.

There have been few attempts to quantify these adverse impacts and it would seem important that such research be conducted at this critical period in the history of the Reserve. Possible studies include those into the types and levels of change caused by visitors and carrying capacity calculations. Reliable data is vital to management planning, helping to identify conflicts and make decisions about resource allocation and protective measures. Information is thus required about entry statistics, visitor origin and method of transport, frequency of visit, size of group, length of stay, movement at the site and behaviour patterns there. Visitor perceptions of their experience and satisfaction with it are another important matter and it would be useful to discover at what point they feel that their enjoyment of the natural surroundings is marred by the sight and activity of others.

However, the positive impacts of the use of the site for recreation should not be overlooked, especially by those in search of evidence to support its continued existence in the face of such strong development pressures. Large numbers are attracted to Bukit Timah, despite limited publicity, and appear to gain from time spent there. The health benefits are of great value in a highly urbanized society like Singapore where many welcome the chance to enjoy public spaces in natural settings. The Reserve also represents an aspect of the country's cultural heritage and it would be unfortunate if it was allowed to disappear like so much of the built heritage has done. Again, this is an area for further research to provide empirical evidence of the contribution of the forest to physical and emotional well-being and its meaning and significance to the local population.

It is already clear that the Reserve serves an educational purpose and provides opportunities for everyone to appreciate and understand nature better. There are excellent links with local schools whereby children are actively involved in a variety of projects while students from Singapore and around the world conduct research studies there. Tropical habitats have great scientific significance and Bukit Timah offers a rare chance to explore how successful remnants of a forest are at maintaining biodiversity.

The impacts discussed above represent both a challenge and opportunity for the management charged with reconciling the many conflicts which arise between conservation and recreation. Although the National Parks Board has a responsibility to conserve areas under its authority, it also has a duty to provide public access to and enjoyment of the various sites; this requires a balancing act which is not easy to perform successfully.

The Management of the Reserve

There is a team of one manager, four assistants and a group of rangers which is responsible for the Reserve with a pool of part-time and unpaid guides who provide additional assistance for tours when required. Ultimate responsibility lies with the Chief Executive Officer of the National Parks Board, but the management team based at the Visitor Centre has a considerable amount of freedom in decision-making. The annual budget is approximately S$300,000 and additional funds are available for specific programmes. Such finance is generally approved provided a sound case can be made for the expenditure, and private sponsors such as the Hong Kong Bank have proved generous.

In terms of policy, the National Parks Board conducted a review of the Reserve when it took over in 1990 and identified areas where use was heaviest, giving rise to negative changes and a need for action. The Board implemented a programme of visitor management through controlled access to footpaths, closing some in order to protect the most fragile locations. In 1992, it opened the Visitor Centre which aims to inform and educate visitors about the Reserve and the need to treat it with respect. The exhibition space helps in the pursuit of these goals, as well as the various leaflets distributed free of charge. Staff work closely with schools and other partners such as the Singapore Environment Council to foster a sense of local ownership and pride in the forest so that residents will be more interested in and committed to its preservation. A scheme was launched in 1997 whereby schools adopt a plot of land already cleared and experiment with plants and saplings, their findings being passed back to the Board for their own planning and planting purposes (Singapore Environment Council, 1997).

Current annual arrivals at the Reserve are an estimated 140,000, although visitors have reached a level as high as 200,000 which has generated anxiety about damage and especially footpath erosion. One solution has been to direct visitors along certain routes and away from others, protecting the most vulnerable by the closure of footpaths or omitting them from maps. The 50-m buffer zone was established with the intention of preventing the encroachment of exotic species and a mountain bike trail created in response to popular demand, the latter actually carefully routed so that it runs largely outside the official reserve and serves as a fire break.

Priorities include securing the buffer zone from development, especially in the north where it is under threat, and maintaining land formally gazetted as protected. The intention is to encourage low-impact activities such as adventure trails and camping and spread the load by possibly opening up the adjoining Central Water Catchment Nature Reserve for leisure purposes. There is an acknowledgement that there will always be competition for land use in Singapore and some forms such as water storage must take precedence, but that compromises can be reached through communication and cooperation amongst interested parties. The staff are proud of their achievements to date and are looking to continue their success in preserving the forest as seen by satellite images of Singapore taken since the 1970s which show the forest becoming greener and thus healthier over the years.

Present and Proposed Strategies

As well as the initiatives described above, the forest reserves of Singapore as a whole are being studied with a view to producing an inventory to act as a database to assist in their proper management. Various parties are involved including the Nature Society, academics, specialist consultants and the National Parks Board itself. Some new species have already been recorded and areas of primary forest discovered (Tan *et al.*, 1996). Another long-term joint research project is being undertaken by the National Institute of Education, part of Nanyang Technological University, and the Centre for Tropical Forest Science at the Smithsonian Institute which is concentrating on the ecological problems facing the tree species; these include stand density, recruitment and regeneration with a long-term inventory in preparation.

The possibility of linking Bukit Timah Nature Reserve to its Central Catchment Area hinterland by way of a corridor that will allow the movement of animals is also under discussion while a scheme is in progress to reintroduce some native animals (*The Straits Times*, 1999). The original intention had been to release as many as 25 wildlife species such as the civet, leopard cat and ant-eating pangolin, but there has been some doubt about the ability of these to survive and possible disruption of the existing ecological balance. After a prolonged study which involved identifying appropriate animals, encouraging them to find their own food and trial periods of release from captivity, the mousedeer was selected for the first experiment. Eight mousedeer from the Singapore Zoo will be released into the Reserve and forest around the MacRitchie reservoir, and their progress monitored for about 2 years using a microchip inserted into the animal's shoulder.

Looking ahead, questions are often raised about whether Bukit Timah will continue to exist and what form it might take in the future. Corlett (1995) identifies the particular problems arising from its size, the extent of exposed 'edge' and separation from other forest areas. The Reserve is long and narrow, crossed by a series of trails and tarmac road, so that at no point is the boundary more than 200 metres away. Conditions on the fringes of the forest in terms of light, temperature and humidity encourage the establishment of non-forest species and speed the extinction of the indigenous species. The Reserve is thus unprotected from the external environment and the internal microclimate disrupted, a difficulty aggravated by its exposed location on Singapore's highest point.

While concluding that further deterioration is inevitable unless action is taken, Corlett remains optimistic that appropriate management can slow this process down and even reverse it – calling for further protection of the margins, planting of open ground, restrictions to access, the removal of undesirable species and the reintroduction of locally extinct species. Such measures are also supported by Wee and Corlett (1986) and Briffet (1990) who argue for a planned management approach to ensure an extended life for the forest. Other proposed measures include the artificial propagation of rare species (Briffet, 1992) and planting of trees which currently are limited in number but contribute significantly to the health of the forest; for example, the fig tree is becoming rare although its presence favours a combination of animal and bird life. Wee (1996), writing about the conditions of ferns in particular, describes the need to reintroduce forest trees which would eventually produce a better canopy to protect the ferns and discourage the invasion of weeds and foreign exotics; together with a surrounding protective border of tall trees, this would allow an appropriate equilibrium to be maintained. Climbers must also be controlled to protect trees from infestation and areas with dead trees could be restored by planting tall forest species.

Additional protection is required for especially vulnerable parts of the Reserve with further closure of paths and strict enforcement of the Nature Reserves Act. There have been some proposals to admit only formally organized groups, led by an experienced guide, properly educated about the site and appropriately sensitive in their behaviour.

The quarry sites within the Reserve remain a problem with a plan to develop them as public parks for recreation. Such parks would act as a means of relieving pressures on Bukit Timah and also meet the recreational demands of many residents, saving unecessary wear and tear of the more delicate parts of the forest. As well as diverting some potential casual users, education and raising awareness levels of others will continue to play a critical role in the future as denying access and excessive restrictions on visitor movement does not appear a viable option. Private sector support and participation is also essential, including sponsorship of research studies.

Future Prospects

The future of Bukit Timah must be considered within the wider context of Singapore and its planning policies for development in general and the environment in particular. The agenda was set in 1991 with the publication of the revised Concept Plan (URA, 1991) by the responsible agency, the Urban Redevelopment Authority or URA; this is the blueprint for Singapore in the next century.

The emphasis of the Concept Plan is on economic growth, but a place is also given to enhancement of the quality of life including the provision of more parks and gardens and the safeguarding of natural heritage. The Concept Plan has been implemented through a series of Development Guide Plans (DGPs) covering 55 planning areas which together represent the new statutory Master Plan; this provides guidelines to landowners about land use and development control, and the most recent revision to the Master Plan was in the late 1990s (URA, 1998).

Bukit Timah Nature Reserve is located in the Bukit Panjang Planning Area where it occupies 46% of the land area. The most recent Planning Report for 1995 (URA, 1995) presents five planning objectives and four planning strategies.

1. Planning objectives:
(a) to optimize residential and other development potential through relocation of incompatible uses;
(b) to preserve and capitalize on the existing terrain, environment and topography of the nature reserve area;
(c) to improve the transport system to facilitate development;
(d) to ensure orderly development and compatibility of uses;
(e) to propose new uses for disused quarries.
2. Planning strategies:
(a) extend the existing Nature Reserve to protect its greenery;
(b) designate areas for parks and recreation;
(c) increase the number and variety of residential dwelling units;
(d) upgrade and develop roads to facilitate development and improve traffic circulation.

It anticipates that 46% of available land will be given over to residential use, 21% to open space and recreation and 17% to roads and infrastructure. Recreational facilities are proposed within and around the Nature Reserve, the vision being to 'create an attractive residential area amidst a green and tranquil setting with good recreational facilities near the Bukit Timah Nature Reserve' (p. 14).

These proposals, currently in the process of implementation, suggest the competition from alternative land uses that the Nature Reserve faces. According to a recent newspaper article (*The Straits Times*, 1998), 'in land-scarce Singapore, flora and fauna do not hold freehold tenure over any spot, even protected nature reserve land. National and strategic interests dictate otherwise. Social and economic needs take precedence and the Government adopts a pragmatic approach to nature conservation'.

The same article goes on to discuss the vulnerability of nature sites to development, despite the presence of the Green Plan, and states that 'as the experience of the nature reserves has shown, its legal status is no guarantee of protection'. It quotes from a paper entitled 'Urbanisation and Nature Conservation' and co-written by the National Parks Board Chief Executive that 'the pressure to regard the nature reserves as a land bank to draw upon for development will intensify ... The fate of nature conservation in Singapore will very much depend upon the political will, which is in turn shaped by the priorities of the people of Singapore'. How much real influence the local population will have on decisions is open to debate and the next section explores possible future scenarios from the perspective of some of the major stakeholders.

Possible Scenarios

Government perspective

The government is committed to the protection and public enjoyment of natural areas and the provision of improved leisure opportunities, claiming that its 'Green and Blue plan safeguards Singapore's natural heritage and also enhances the ecology of its urban environment' (URA, 1991: 31). It also recognizes that remaining pockets of relatively untouched landscape are threatened by easy access and heavy usage, and should provide a physical challenge for those visiting as well as retain their naturalness and authenticity. In pursuit of such goals, the authorities describe a programme of capitalizing on the country's natural assets with the establishment of adventure and nature parks at five sites including Bukit Timah.

These statements and proposals would suggest that there might be some changes within the Reserve while the Concept Plan also notes that by the Year X, the date when Singapore's population is expected to reach the critical level of 4 million, development will have to be expanded into currently unpopulated areas. It was originally expected that Year X would be reached some time after 2010, although recent reports indicate that the census in 2000 will record nearly 4 million residents. The projections of population growth confirm the importance of the land issue in any future planning and development decisions.

As the URA's head of strategic planning noted (*The Straits Times*, 1998), Singapore occupies about the same area as a median size city in a developed nation at 645 km^2; however, these cities do not have to worry about space for military training or water catchment. These preoccupations were expressed by the head of local planning in the same article who said, 'future development and reclamation projects affecting some nature areas cannot be entirely precluded given that there could be other more important competing uses for such land, particularly if they are of national or strategic interest, for example housing and infrastructure needs'. The Environment Minister has also stated that land constraints will make it very difficult to set aside land for nature appreciation alone, recommending that environmental groups organize educational gateways that provide opportunities for Singaporeans to enjoy nature beyond the Republic.

Taking into account these official views and government policy, it would appear that Bukit Timah is not guaranteed protection from further encroachment and could well be further reduced in size and altered in character. It is unlikely that it will disappear completely, but may well move from being a tropical forest of international and national significance which offers a sense of wilderness to an urban park with inescapable evidence of interference, management and man-made facilities.

Tourist industry perspective

The Reserve is already promoted as an attraction of Singapore in the official guide with visitors invited to 'take a walk on the wild side' (STB, 1999: 32). It is described as sharing with Rio de Janeiro the distinction of being one of the only two rainforests in the world within city boundaries, offering a fascinating array of plant, animal and bird life.

Although overseas tourists do not make up a large proportion of visitors, partly because Singapore remains a short-stay destination of 3–4 days allowing little opportunity to explore the island and its less well-known and accessible attractions, Bukit Timah would seem to be of growing appeal given the increased popularity of and enthusiasm for all forms of nature tourism. These trends have been recognized by the Singapore Tourism Board or STB (formerly the Singapore Tourist Promotion Board or STPB) which listed the theme of Nature Trail amongst a total of 11 appropriate to promote in its Tourism 21 strategy (STPB, 1995), designed to direct the tourism industry and shape its evolution in the new millennium.

The theme includes Bukit Timah Reserve as an 'opportunity area' with a 'story' of back to nature and 'possible activity cluster' of ecotours such as bird watching, general nature walks, hikes and tropical forest experience in addition to environmentally friendly merchandising at visitor centres. The Board set a 5-year timeframe for the realization of these objectives and identified the National Parks Board as the lead agency and nature groups and societies, relevant university departments and travel agents as other implementors besides itself.

Progress towards putting the Tourism 21 strategy in place has been delayed by the Asian economic crisis which commenced in 1997 and has had a serious effect on tourist arrivals and spending, but it remains the formal expression of tourism policy and the proposals clearly have some bearing on the future of Bukit Timah. Although tourism interest and expenditure might be seen as a strong argument for the preservation of the Reserve, its successful promotion and the arrival of more overseas visitors would place further pressure on the resources with a possible demand for additional facilities there such as coach parking, toilets and catering. Careful visitor management will be required to minimize the impacts and ensure that a balance is reached between the demands of the tourists, the tourism industry, the local community and the forest itself.

Bukit Timah might also become better known amongst both visitors and residents as a result of a new map being produced by the Green Volunteer Network, a group set up by the Singapore Environment Council to stimulate greater awareness of environmental issues and encourage participation. The map will chart all of Singapore's 'green' areas and the group would like to see it published by the Tourism Board and available to tourists in order to help them explore the island more thoroughly.

Conservationist perspective

As previously stated, conservationists have taken a very active interest in Bukit Timah since its early days and sought to draw attention to its vulnerability and the constant threat of degazetting and development. The size of the Reserve and its shape make it especially vulnerable to the influences of the external environment and limit the number of species supported. One estimate for the future is that only 20% of Singapore's original plants and animals will find a home in the Reserve and Table 2.1 presents the worst-case scenario for species loss (Lum and Sharp, 1966: 102).

The authors maintain, however, that despite such a possible decline in tree species the primary forest appears to be renewing itself successfully and even the disturbed parts of the forest are returning to their original character. Others might be rather more pessimistic and suggest that while the picture may not be entirely hopeless, a programme of active management is essential if the Reserve is to retain at least its current state of health; without this, further deterioration becomes inevitable.

Table 2.1. An estimate of Singapore's possible species survival at Bukit Timah.

Organism	Original forest species in all Singapore	Worst-case scenario for Bukit Timah's future
Birds	268	52
Mammals	79	15
Reptiles	112	22
Amphibians	25	5
Trees	800	154
Ferns	170	33

Visitor perspective

As already recorded, little research appears to have been conducted on the subject of visitors to the site or attitudes amongst the local community towards its preservation. Increasing numbers indicate its growing popularity and the stresses of everyday living in Singapore underline its importance as a means of escape and release, while it offers an appealing contrast with most of the rest of the attractions on offer for the overseas visitor. There would thus appear to be strong demand for the forest as a recreational resource, and it has acquired an additional symbolic significance for some. The view is expressed strongly by Lum and Sharp (1996) who write '... Bukit Timah transcends its function as a place to commune with nature, to invigorate a desk-bound body, or to describe a new species of animal. Bukit Timah is a symbol, a manifestation of our collective national psyche, of our commitment to making this city state a better place. If we cannot preserve and nurture ... a forest rich in biodiversity and steeped in scientific tradition, how far have we progressed as a society? ... It is here at Bukit Timah that we come full circle, for the Reserve is both a link to our past and a harbinger of our future' (p. 113).

Even if such calls as these meet with a sympathetic response from the appropriate authorities, the question remains of whether a rising number of visitors will be satisfied with their experience of the forest given the constraints of its size and capacity. Restrictions to access and more controlled visitor management might be necessary to avoid the situation of an unacceptably high level of visitors and the resultant stress that they will impose.

Conclusions

Each party therefore has a particular interest in Bukit Timah and a different set of priorities. For the Reserve to survive and prosper, all groups will have to contribute in some way. Government will need to continue to provide protection from development to the forest and surrounding buffer zone and the tourist industry be prepared to act responsibly in its promotion of it as an attraction, encouraging visits during quieter periods of small groups led by experienced and knowledgeable guides. Conservationists and scientists must pursue their studies, with international collaborative action when necessary, to improve understanding of the forest's biodiversity and experiment with techniques of recovery and repair to assist in its effective management – supported by sponsorship from private industry and drawing lessons from overseas experience. Visitors might have to accept more limited access, adhere strictly to a code of conduct and possibly contribute financially to the maintenance of the Reserve with the introduction of entrance fees which might also act as a control on numbers.

Within the forest, existing visitor management policies should continue using the footpath network to direct movement and flows, and closing off especially sensitive parts. Monitoring and evaluation of policies should be a continuing process. The Visitor Centre and team of Rangers have an important education and policing function to fill. Visitors could be encouraged to use alternative, less vulnerable, parks depending upon their particular needs with the introduction of restrictions on admissions during the busiest times. The forest could even be closed for short durations to allow time for rest and recovery. Practical constraints do exist, however, and successfully implementing some of these schemes remains a problem given the large boundary of the forest which offers several points of unofficial entry, and possible popular resistance.

However, the debate about land use overshadows the future of Bukit Timah. Developers argue that housing and infrastructure needs must take priority and government speaks of land constraints because of the space required for military training and water catchment, suggesting that there are limits to the amount of the island which can be devoted to nature conservation. At the same time, conservationists are worried about the growing pressures on the country's natural heritage, the effectiveness of the Green Plan and the loss of indigenous species whilst more visitors are seeking out the pleasures of the Reserve. Bukit Timah is caught in the middle of this debate and its survival as a unique urban rainforest will require effective internal management and a commitment amongst relevant external parties.

References

Briffet, C. (ed.) (1990) *Masterplan for the Conservation of Nature in Singapore.* Malayan Nature Society, Singapore.

Briffet, C. (1992) The potential for the conservation of nature in Singapore. In: Chua Beng Huat and Edwards, N. (eds) *Public Space: Design, Use and Management.* Singapore University Press, Singapore, pp. 115–127.

Briffet, C. and Sim Loo Lee (eds) (1993) *Environmental Issues in Development and Conservation.* SNP Publishers, Singapore.

Chua Ee Kiam (1993) *Nature in Singapore: Ours to Protect.* Landmark Books, Singapore.

Corlett, R.T. (1992) Conserving the natural flora and fauna in Singapore. In: Chua Beng Huat and Edwards, N. (eds) *Public Space: Design, Use and Management.* Singapore University Press, Singapore, pp. 128–137.

Corlett, R.T. (1995) Introduction. In: Chin See Chung, Corlett, R.T., Wee, Y.C. and Geh, S.Y. (eds) *Rain Forest in the City: Bukit Timah Nature Reserve.* Gardens' Bulletin Singapore, Supplement No.3. National Parks Board, Singapore.

D'Silva, E. and Appanah, S. (1993) *Forestry Management for Sustainable Development.* Economic Development Institute of the World Bank, Washington, DC.

Edwards, D.S., Booth, W.E. and Choy, S.C. (1996) *Tropical Rainforest Research – Current Issues: Proceedings of the Conference held in Bandar Seri Begawan, April 1993.* Kluwer Academic Publishers, London.

Hunter, C. and Green, H. (1995) *Tourism and the Environment: A Sustainable Relationship?* Routledge, London.

Lum, S. and Sharp, I. (eds) (1996) *A View from the Summit: The Story of Bukit Timah Nature Reserve.* Nanyang Technological University and the National University of Singapore, Singapore.

Mathieson, A. and Wall, G. (1982) *Tourism: Economic, Physical and Social Impacts.* Longman, London.

National Parks Board (1999) *National Parks Board: Making Singapore Our Garden.* Website. http://home1.pacific.net.sg/~nparks.

National Parks Board (undated) *Bukit Timah Nature Reserve.* National Parks Board, Singapore, leaflet.

Sharma, N.P. (1992) *Managing the World's Forests: Looking for Balance between Conservation and Development.* Kendall/Hunt Publishing Company, Iowa.

Singapore Environment Council (1997) *Greening a Corner of Singapore: A Pilot Study of Reforestation Techniques in Bukit Timah Nature Reserve.* Singapore Environment Council, Singapore.

Singapore Science Centre (1995) *A Guide To The Bukit Timah Nature Reserve.* Singapore Science Centre in collaboration with the Nature Reserves Board, Singapore.

Singapore Tourism Board (1999) *Official Guide: Singapore New Asia.* Singapore Tourism Board, Singapore.

Singapore Tourist Promotion Board (1995) *Tourism 21: Vision of a Tourism Capital.* Singapore Tourist Promotion Board, Singapore.

The Straits Times (1996) *Singapore: Nature's New Neighbours – Forests on One Side, Condos on the Other,* 27 September.

The Straits Times (1998) *Can Land-scarce S'pore Afford Nature Conservation?,* 13 September.

The Straits Times (1999) *Mousedeer to be Released into Reserves,* 6 April.

Tan, Wee Kiat, Lee Sing Kong, Wee Yeow Chin and Foong Thai Wu (1996) Urbanisation and nature conservation. In: Ooi Giok Ling (ed.) *Environment and the City: Sharing Singapore's Experiences and Future Challenges.* Times Academic Press, Singapore, pp. 185–199.

Urban Redevelopment Authority (1991) *Living the Next Lap: Towards a Tropical City of Excellence. The Concept Plan.* Urban Redevelopment Authority, Singapore.

Urban Redevelopment Authority (1995) *Bukit Panjang Planning Area: Planning Report 1995.* Urban Redevelopment Authority, Singapore.

Urban Redevelopment Authority (1998) *The Planning Act 1998: Master Plan Written Statement 1998.* Urban Redevelopment Authority, Singapore.

Wee Yeow Chin (1996) Ferns as crisis indicators. In: Lum, S. and Sharp, I. (eds) *A View from the Summit: The Story of Bukit Timah Nature Reserve.* Nanyang Technological University and the National University of Singapore, Singapore, pp. 96–97.

Wee Yeow Chin and Corlett, R. (1986) *The City and the Forest: Plant Life in Urban Singapore.* Singapore University Press, Singapore.

3

Competing Interests on a Former Military Training Area: a Case from Estonia

Mart Reimann and Hannes Palang

Introduction

The former Aegviidu military training area (330 km^2) lies 50–60 km east of Tallinn, the capital of Estonia. For some forty years it has been strictly closed to the public. The Soviet army used and damaged less than 10% of the area, leaving the rest untouched. In 1991, after the military had left, the area was taken under protection as the Põhja-Kõrvemaa Landscape Reserve.

The Reserve has a diverse natural landscape (post-glacial relief, old-stand forest, massive virgin mires and numerous clean lakes) and forms a habitat for many rare and endangered animal and plant species.

In addition to nature conservation, there are three more sectors interested in using the area. Firstly, after the Soviet military left the area, the number of visitors keeps growing, in some places already exceeding the carrying capacity of the area. Due to its proximity to the capital it is perhaps the most visited protected area in the country.

Secondly, the area is still unpopulated, but the land reform aiming at re-privatizing the land to their former owners has already brought back some people, while other new owners are interested in cutting the trees and selling them. Thirdly, the Estonian military want to re-establish a training ground here.

The paper concentrates on how these often conflicting interests could be managed in the reserve. Public investigations of visitors, landowners and forestry employees give an insight into the situation in the area. These, together with the basic considerations of nature conservation and forest management should lead to finding possibilities for sustainable management of the area so that both nature conservation and recreation could be continued in the area.

Nature Conservation

Nature conservation authorities were the first to discover the area abandoned by the Soviet military. Due to the high natural values, a quick decision was made to take the area under protection.

Having the area under nature conservation has several advantages. It guarantees that the natural values will be preserved. This is done though the protection rules that list what kind of activities are allowed in the reserve. Furthermore, every protected area should have a management plan that shows in detail what should be done to maintain the values. As the protection rules and management plan have legal power, they are obligatory to other stakeholders, such as developers and planners. This approach ensures that nature conservation interests will have the highest priority in the area and all other interests will have to comply with these.

Also, recreation benefits from the protection regime. People come here because of the natural values, and until these are maintained, recreation could continue. A protection regime also helps to keep the number of visitors within the limits of the carrying capacity of the area.

The only negative impact of the protection regime could be the limits to forest management that might decrease their profits.

Natural values

Põhja-Kõrvemaa is a part of the large forest and mires belt called Middle Estonia, which is approximately 40 km wide and runs from north to southeast Estonia. It is a geobotanically important region because a number of plant species grow here on the edge of their distribution area and several rare and endangered species are present. However, due to its limited access during the Soviet times, the natural values are not always well documented; Table 3.1 is based on the works of Mäemets (1977), Tõnisson (1991), Arhus County (1995), Karofeld (1995) and Kukk (1996).

The area is rich in post-glacial formations, which are traversed by series of long ridges of eskers and bulge-like elevations (relative height 35 m and steeper slopes reaching 30°), lying athwart the terminal moraines of the continental ice. The highest point is 97 m above sea level.

More than 30 larger lakes are found in the Reserve, plus hundreds of bog pools. Many lakes have very clean water and rare water plants. The main rivers in the area are the clean meandering Soodla and the Valgejõgi on the eastern border. The Soodla water reservoir supplying Tallinn with drinking water has been created on the Soodla River. As the Soodla flows slowly in a natural, not dredged riverbed, natural flood plains and mires, relatively rare in Europe, are found here.

Variations in the terrain combined with great differences in soil fertility and water availability have resulted in a very high diversity of forest and veg-

Table 3.1. Main plant and animal species of the Põhja Kõrvemaa Landscape Reserve.

Species	Habitat
Water lobelia (*Lobelia dortmana*)	Lake Mähuste
Quillwort (*Isoethes echinospora*)	Lakes Mähuste, Paukjärv, Jussi Linajärv
Small water lily (*Nuphar pumila*)	Several lakes
Burreed (*Sparganium augustifolium*)	Several lakes
Russian yellow oxytropis (*Oxytropis sordida*)	Western part of the reserve (the only place in Estonia)
Pulsatilla patens	Dry sandy soils on heaths and in pine forests
19 species of orchids (*Orchidaceae*)	Fens
Brown bear (*Ursus arctos*)	
Wolf (*Canis lupus*)	
Lynx (*Lynx lynx*)	Forests
Elk (*Alces alces*)	
Roe deer (*Capreolus capreolus*)	
Wild boar (*Sus scrofa*)	
Flying squirrel (*Pteromys volans*)	Aspen forests
European mink (*Mustela lutreola*)	River Valgejögi
Beaver (*Castor fiber*)	Rivers and streams
Otter (*Lutra lutra*)	Rivers and streams
Muskrat (*Ondatra zibethicus*)	Rivers and streams
Crane (*Grus grus*)	Bogs
Black stork (*Ciconia nigra*)	Forests
Golden eagle (*Aquila chrysaetos*)	Bogs and forests
Capercaillie (*Tetrao urogallus*)	Bogs and forests
Black grouse (*Tetrao tetrix*)	Bogs and forests
Warblers (Silviinae)	Forests and grasslands
Woodpeckers (Picidae)	Forests
Owls (Strigiformes)	
Ducks (Anatidae)	Rivers and lakes
Perch (*Perca fluviatilis*)	Rivers and lakes
Crucian carp (*Carrassius carrassius*)	Rivers and lakes
Roach (*Rutilus rutilus*)	Rivers and lakes
Pike (*Esox lucius*)	Rivers and lakes
Tench (*Tinca tinca*)	Rivers and lakes
Crayfish (*Astacus astacus*)	Rivers and lakes

etation types in the Reserve. About 40% of the Reserve is covered by forests, 50% wetlands and only 10% by man-made open areas. The dominating species in forests is Scots pine (*Pinus sylvestris*), frequently mixed with Norway spruce (*Picea abies*) and birch (*Betula* sp.). In the south of the Reserve, the forest type changes gradually into the Norway spruce dominated forest covering most of central Estonia.

Mires and raised bogs that cover large areas are among the most valuable biotopes in the Reserve. The paludification in Põhja-Kõrvemaa started some 9–10,000 years ago. The biggest mires in the area are Koitjärve bog (1750 ha, peat depth 7.4 m), Kõnnu Suursoo bog (1620 ha, 6.3 m) and Võhma bog (870 ha, 6.5m). The last two *Sphagnum*-dominated bogs in particular, with large numbers of bog pools, are in a virgin state and serve as refuges for several plant and animal species. A eutrophic floodplain mire, which is over-flooded in spring, lies in the Soodla River valley.

The nature conservation regime of the area

In 1991 the Soviet army left the Põhja-Kõrvemaa area. At that time the majority of the forest had been largely undisturbed for about 40 years. During these years the natural succession formed a forest close to a natural forest with stands composed of a varied mixture of young, old and dead trees of different species. It was a forest with low commercial and high ecological value.

On 29 October 1991, the Põhja-Kõrvemaa Nature Reserve was established in the (former) military area by the Harju County Council. Thus the old idea of Jakob Ploompuu, a local book publisher who had already in the 1920s called for the creation of a protected area, had come true (Tõnisson, 1984).

Nature conservation in Estonia is based on the Act on Protected Natural Objects (APNO), adopted by Parliament on 9 July 1994. According to this, nature conservation is carried out on a state level. Protected areas are divided into four categories: national parks, nature reserves, landscape reserves and programme areas. The management of these areas takes place on the basis of protection rules approved by the government via legal restrictions and obligations that are established for separate zones. Depending on the specific features of the protected area concerned, protection rules may alleviate the restrictions and obligations stipulated by APNO.

According to APNO, protected areas are divided into three zones: strict nature reserves, special management zones and limited management zones. In a strict nature reserve, economic activities and human presence are prohibited. Only enforcement, scientific and rescue activities are allowed. Strict nature reserves can exist only in national parks and nature reserves. In the IUCN system, strict nature reserves correspond to category Ia (IUCN, 1994).

The objective of the special management zone is to support the natural development of ecosystems and, consequently, minimize human impact. Human presence is permanently or temporarily permitted. Economic activity is only allowed for nature conservation or recreation objectives (managing semi-natural biotopes, erecting a watchtower or cutting some trees for better view). Special management zones may occur in national parks, nature reserves and landscape reserves.

A limited management zone comprises areas of a quite different degree of human impact. Profitable economic activities can be continued in a nature friendly way. All kinds of economic activities not prohibited by law or the protection rules are permitted here. A limited management zone can occur in all of the protected areas. In the IUCN system a limited management zone corresponds to Category V (IUCN, 1994). Generally the share of the limited management zone is largest in landscape reserves and smallest in nature reserves (Ministry of the Environment, 1998).

The new protection rule for the Põhja-Kõrvemaa Landscape Reserve was approved by the government on 26 August 1997. It divides the Põhja-Kõrvemaa Landscape Reserve into two special management zones (altogether 59 km^2) and one limited management zone (70 km^2). Harju County Government is responsible for managing the area while the Koitjärve Forest District carries out local administration.

Forestry and Land Ownership

Forestry has been considered the main economic possibility of the region. Fortunately for nature, the population of the area is extremely low. Still, the possibility of massive logging could hinder both nature conservation and the recreation value of the area.

Population history

Human presence has been low over the years in Põhja-Kõrvemaa. In general, the soils are poor, badly drained and not very suitable for agriculture. As a result, the average population density before the Soviets arrived in the 1950s was only one person per km^2. Still, Põhja-Kõrvemaa has an important place in Estonian cultural history. Many writers and actors used to spend their summer vacation here. Anton Hansen Tammsaare, one of the most important writers in Estonian history, lived in Koitjärve in 1911–1918. Every Estonian knows his descriptions of the beautiful nature of Põhja-Kõrvemaa and the local villagers.

In 1953 Põhja-Kõrvemaa became part of a large Soviet military training area (330 km^2) and the civil inhabitants had to leave the area. At that time more than 30 small farms and ranger houses existed there. Traces of these can still be found today, such as the ruins of former houses and grasslands that were once fields.

The Soviet army used the area for air and artillery bombing, tank and infantry practice. A mock-up of a nuclear bomb was dropped on the area. Also, experiments of how to use laser weapons against tanks were carried out here (Liim, 1997), but army activities were mainly concentrated in small areas and most of Põhja-Kõrvemaa was left undisturbed. The largest damage is in the Jussi impact area (about 1.5 km^2), an area used for artillery bombing in the central part of the territory, just west of the Jussi lakes. Today as the result of the bombing and fires the area is without forest and mainly dominated by heather (*Calluna vulgaris*). This area looks almost like tundra.

Ownership changes

Currently, 90% of the Põhja-Kõrvemaa Landscape Reserve is owned by the state. After regaining independence in 1991, the Estonian government decided to give the once nationalized property back to the former owners or their descendants. Since then, around 25 former farmers have got their land back or have applied for it. However, it is not certain yet how large the areas they will actually receive will be.

According to the Act on Protected Natural Objects the economic purpose or use of land within a protected area cannot be changed without permission from the authorities. The same law makes it possible to compensate for losses resulting from protection by reduced taxation. Owners whose land is located within strict nature reserves or special management zones may request the state to purchase the land in question or provide a substitute. In Põhja-Kõrvemaa, the lands of three farms, which are to be given back to owners, are located in special management zones. The rest of the farms are all situated in the limited management zone.

Currently, there is no electric supply on the territory of the Reserve and the road conditions are rather bad. At present, only one owner lives in the middle of the area. Few farms are near the Tallinn–Narva and Tallinn–Aegviidu roads, which border the area in north and south. An investigation among the owners, potential owners, local communes and forest district employees (Reimann, 1996) has shown that 70% of them want to use their land themselves; two of them want to live here permanently, the rest want only to use the forest or build a summer cottage. None of the owners wants to use the land for agricultural purposes. However, the investigation was carried out almost 4 years ago and there are no signs that any of the owners has started to build. Building or management needs investment, but half of the owners have an income lower than average while the average age of owners is 61 years. Therefore it is hard to believe that large-scale activities will be undertaken in the near future.

Forestry

The Soviet army changed little in Põhja-Kõrvemaa – only small areas along the main roads have been cut clear. During the military period the forest was considered to be a Soviet forest and the authorities in Moscow managed it according to the Soviet military forest law. Management of other forest areas in Estonia remained under the principles of the Estonian forest law.

In 1993 the new Forest Law came into force. Since then many transformations have happened in forestry management structure and amendments have been made in the Forest Law as well. According to the Act on Protected Natural Objects, forests within special management zones and limited management zones fall into a category of either preservation or protection forest in the Forest Law. There are some contradictions between these two laws. Forestry activities can be banned according to the APNO, while forestry activities are allowed with some limitations according to the Forest Law. This results in conflicts between nature conservation authorities and more production-minded forestry authorities. The latter often dictate the timber management situation in protected areas.

Since 1993, the Koitjärve Forest District has been responsible for the forestry operations carried out on the state-owned part of the reserve. According to the protection rules, timber harvesting is banned in the special management zones and final cutting is banned in limited management zones. The district itself is governed by the Northeast Estonian Forest Management Region, which is the largest in the country. Compared to its total production the share of the Koitjärve District is so small that after discussions between the county government and the Forest Management Region authorities a decision was made that there was no need to cut the forest. This means that Põhja-Kõrvemaa is the first large protected area in Estonia where no profit-oriented timber production activities are carried out.

Military Activities

Since the formation of the Estonian army in 1991, it has shown great interest in continuing the use of Põhja-Kõrvemaa for military purposes. After the Soviet army had left Põhja-Kõrvemaa, the Estonian army temporarily carried out some infantry practice in the area. At that time many defence forces authorities, often former Soviet officers, were of opinion that the Estonian army must possess all former Soviet military installations. Initially the Estonian army damaged nature in the same way the Soviet army had. This caused serious conflicts with the Koitjärve Forest District. In 1996 the Estonian army wanted to create a central military training ground on the area of the Põhja-Kõrvemaa Landscape Reserve. However, an environmental impact assessment demonstrated that it was impossible to create a large-scale military training area within a landscape reserve.

Currently the army is interested in the area bordering the Reserve in the east. The problem is that the military training area is planned close to the most valuable parts of the Reserve. The army also wants to set up more than 15 artillery positions within the Reserve.

Recreation

Recreation is the most popular activity in the area. However, as it has been declared a landscape reserve, visitors should comply with the protection regime. It may easily happen that the amount of visitors or their behaviour exceeds the carrying capacity of the area. There are already places where the natural values have been damaged. On the other hand, in certain places and at certain times the number of visitors already exceeds the social carrying capacity. So recreational activities should be organized so that the most vulnerable areas could stay intact and the flow of visitors managed in a way that they would not disturb each other.

Facilities

Besides nature protection, one of the purposes of the Reserve is to allow people to experience undisturbed nature and scenic landscapes and to promote the knowledge of nature. The first priority goal of the Estonian National Environmental Strategy (Ministry of the Environment, 1997) is to promote environmental awareness. The Põhja-Kõrvemaa Landscape Reserve provides good possibilities for nature education because of the diverse nature values, its closeness and good connection by train and road to the capital Tallinn (where one third of Estonia's population lives).

The main recreational activities on the territory of the Reserve are picking berries, fishing, hiking and picnicking.

Greatest attention has been paid to creating hiking facilities. Currently there are three nature trails marked in the forest for which brochures are also available. They are easily accessible and support the purposes of nature education. For more adventurous people, a 36-km long trail has been marked that crosses the territory of the Reserve and includes a camping night in the wilderness.

Despite all efforts, the main attractions for visitors are still mushrooms and berries. They can be found almost everywhere in Põhja-Kõrvemaa. The season starts in June with blueberries (*Vaccinium myrtillus*) and mushrooms, continues with cowberries (*Vaccinium vitis-idea*) and ends in October with cranberries (*Vaccinium oxycoccus*).

Several good fishing grounds exist within the Reserve. Everybody is allowed to fish with a fishing rod without a reel. Fishing with a reel is regulated by protection rules and a licence issued by the county authority is required. A special activity is the night-time crayfish catching in August and September.

According to the protection rules, camping and fires are only allowed in officially provided places. There are five authorized camping places and six fireplaces in the area. However, the existing camping places are not sufficient for the number of visitors. People look for silence and privacy and if one group is already camping in a camping place the other group will find another. If no official camping place is available people camp where it is not allowed. Camping is often connected with parties. In official camping places there are waste bins but in illegal camping places litter, such as empty vodka bottles, cans and packing, is often found.

Visitors

Since the Soviet army left, information about this large wild and beautiful area has been passed from person to person and the number of visitors has been increasing year by year. Investigations in July and August of the last 2 years have shown that at summer weekends more than 200 cars and 700 persons visit the area daily. In the rest of the week the respective numbers are 60 cars and around 200 persons (Reimann, 1998). Most of the visitors come by car, just a few by bicycle or on foot. At the weekends around 80% and during the week 50% of visitors come from Tallinn.

Although the landscape is particularly suitable for all kinds of hiking trips, only 5% of the visitors come with this aim. The most popular activity among the visitors is picking berries and mushrooms (68% of all visitors).

Another 14% of all visitors come to have a picnic and many of them stay overnight in tents. Ten per cent of all visitors camp and make fires. Most of them are picnickers.

Nine per cent of visitors come to fish and 4% to catch crayfish. Fishermen come usually alone, except for crayfish catchers who form groups and have a kind of ceremony.

More than half of the visitors are ethnic Russians who are interested mainly in the utilization values of nature. Ethnic Estonians are more concerned with the aesthetic values of nature. Hikers and crayfish catchers as well as picnickers are usually Estonians. The majority of berry-pickers and fishermen are Russians. Among the Russians the fishermen tend to be the only ones who really care about nature. They find conflicts with the berry-pickers who make noise and leave litter, while they themselves try to keep quiet, make their campfires in official places and place their litter in trashcans or bury it.

The most visited places in the area are the Järvi lakes, which are the most accessible ones, in the northern part of the area. These lie in the limited management zone and do not have a very high nature conservation value, but many visitors go to the Jussi lakes and Lake Paukjärv that fall within the special management zone because of their clean water and rare plants.

Conflict areas

Today the lakes in the central part of the area suffer from too high recreational pressure. The shores of the cleanest and most valuable lakes are usually sandy and with very fragile vegetation. In many places around Lake Paukjärv and the Jussi lakes the undergrowth of the pine forest has been damaged. Too intensive recreation can be dangerous also to the water of lakes. Some people drive their cars very close to the lakes or even wash their cars in lake water. Lake Mähuste is a special case, as it lies in the limited management zone in private land, but has the biggest number of rare plant species compared to the other lakes in the area (Tõnisson, 1991; Kukk, 1996). Therefore the lake should be kept out of the reach of ordinary visitors.

The New Management Plan

Due to the protection regime, management of the area is to be done from the top down, i.e. all local initiatives should comply with the protection rules. In this case, the aim has been to fit recreational interest better with the nature conservation interests. Still, the management plan will have to consider all parties and their interests.

Administration and nature conservation

Among the first steps for the new management would be to create a new administration for the Reserve. Currently this task is carried out by the Forest District, but the county government has planned to employ two persons who would be responsible for managing the Reserve. They will also have to implement most of the planned actions included in the management plan of the reserve.

Nature conservation activities will be carried out according to the existing protection rules. As nature conservation will remain the priority in the area, these rules will also be the basis for solving conflicts with other interest groups. Usually the biggest conflict occurs with forestry and private owners. In Põhja-Kõrvemaa the biggest conflict is with recreation. Nature protection conflicts with the military cannot be compared with other cases in Estonia because in some other protected areas there may only be small-scale infantry training.

Overall, there is not enough knowledge about the nature of Põhja-Kõrvemaa and more research is needed. The protection rules contain mostly restrictions and limitations for economical activities. In the near future a management plan for the Põhja-Kõrvemaa Landscape Reserve will be worked out that will include detailed zoning for recreational activities. Other priorities also include creating more recreational facilities, limiting car traffic and improving conditions for hikers and bikers in the area. The purpose is the better use of the potential of Põhja-Kõrvemaa and to make it an important site of nature education in addition to its use for picnickers and berry-pickers.

Recreation

Of the four conflicting interests, recreation is the most difficult to regulate. Investigation among the owners showed that the majority of respondents thought that recreation should be reasonably limited in the area (Reimann, 1996). There are two possible ways to do this. Firstly, creating physical barriers such as bars, stones or ditches can impede access to the area by car. Another possibility is to promote activities in less sensitive areas. Three years ago the Forest District tried to close the bridge across the Soodla River to keep cars away from Lake Paukjärv and the Jussi lakes. Bars have been erected many times, but even metal ones got destroyed during the following nights. In 1997 wooden sticks were put around the lakes to prevent cars getting too close to the water. This has worked better but still there are often missing sticks that need replacing. It seems that many visitors do not want to be regulated or restricted.

The main efforts for regulating recreation in the Põhja-Kõrvemaa Landscape Reserve should concentrate on promoting activities in the less sensitive areas in the limited management zone of the reserve. This has started with opening views and erecting viewtowers. Also two big camping sites have been created which can accommodate bigger groups and can be accessed by coaches as well. In the south-west, close to the Tallinn–Aegviidu road, a nature trail was marked. Also, visitors are directed to trails outside the Reserve, south of Aegviidu.

Bad road conditions are one of the factors that keep people away from the area. Also, some extra camping and fireplaces are badly needed. Brochures and maps about the Reserve have been published, but according to polls, this is not enough. Interest in these was mainly expressed by Estonian-speaking visitors, while Russian-speakers said that the only information they need is where the best berry and mushroom places are. One explanation for this could be that the brochures have been published in Estonian and this remains a language barrier for the Russians. Viewtowers are considered necessary by some, but by others a waste of money and good only for hunters.

Hiking trails should be more popular. In the future the central part of the Reserve will be accessible by foot or by bicycle and preferably through the nature trails. This is also the wish of the landowners (Reimann, 1996). As there are many people who want to learn about nature, guided tours are thought necessary. Seventy-five per cent of landowners thought that organized nature tourism should certainly be developed in the Reserve and the main visiting style should be in guided groups. Some guided tours have already been organized by the County Government and the Forestry District.

In 1999, the building of the Visitors' Centre will start in the southern, most accessible part of the Reserve. It will have a permanent exhibition about the Landscape Reserve and its nature. It will also work as a base for a nature school where pupils and tourists can go for 1- or 2-day trips. As the nature trail departs there, guides will be available to show people around the area. There will be a library and sales of publications about the reserve, and also a small hostel and a refreshment shop.

Finally, eight information boards will be set up with maps and behaviour rules in the Reserve.

Forestry

Right now there is no direct threat from the forest authorities. However, the future will depend on the economical situation of the Northeast Forest Management Region and on the situation of the timber market. With the change of administration, some people in the forest district may lose their jobs, as the Forest District will be dissolved or joined to another forest district.

Another issue is related to the private owners. In the limited management zone some limited cutting could be done in the future, as there are mainly pine forests where cutting does not harm nature conservation and recreational values as much as in other forest types. Selective cutting could even improve the recreational value of a pine forest.

Military activities

The military use of the Reserve can cause conflicts with all the other activities. The military training area is planned just next to the eastern border of the Reserve where the best virgin-like forests are situated. The planned impact area is only 1 km away from the Jussi lakes, a popular recreational area. Research has shown that in former training areas shell splinters of bombs have damaged trees within a radius of more than 500 m around the impact areas (Arukaevu and Pregel, 1994). The impact area is also very close to the nest of a golden eagle. The situation would improve if the impact area were shifted more to the east. Some of the many artillery positions that are planned inside the reserve are in fragile places that could be damaged while setting up the guns. The problem will become more delicate if the training area is also to be used by NATO forces.

Conclusions

There are four competing interests using the Reserve. Firstly, nature conservation authorities want to keep the high natural values of the area as intact as possible. Secondly, the former owners and their heirs intend to make as much profit out of their lands as possible, and most often the best way is thought to be logging. Thirdly, as the area has almost no human population, the military think the area should be used as a military training area. Finally, the area has become famous as a recreation site among the inhabitants of Tallinn. As the area has been taken under protection as a landscape reserve, it is supposed to have protection rules as well as a management plan. The new management plan tries to find a compromise between all these interests. The prevailing opinion, shared by the nature conservation authorities as well as local communities is that logging and military activities should be kept away from the area. As the area belongs to a landscape reserve, nature conservation will remain the first priority. Efforts have been made to keep tourism, recreational and forestry activities away from areas with the highest ecological values. The plan tries to give a second priority to recreation, especially so that the scientific and educational values of the area could be better realized. The new Visitors' Centre and nature trails also encourage this. Forestry works should be kept as limited as possible, but still they cannot be avoided. Problems with the military are the most delicate ones and are not that easy to tackle, but hopefully even here a compromise could be found.

References

Arhus County (1995) *North-Kõrvemaa Nature Reserve. A Conservation Plan for an Estonian Nature Reserve.* Arhus County, Denmark.

Arukaevu, M. and Pregel, P. (1994) Koitjärve metskonna metsad. BSc thesis at the Estonian Agricultural University, Tartu, Estonia.

IUCN (1994) *Parks for Life: Action for Protected Areas in Europe.* IUCN, Gland and Cambridge.

Karofeld, E. (1995) Põhja-Kõrvemaa looduskaitseala arengustrateegiast. Manuscript at the Institute of Ecology, Tallinn.

Kukk, Ü. (1996) Põhja-Kõrvemaa floora haruldaste taimeliikide seisund. Manuscript at the Harju County Government, Tallinn.

Liim, J. (1997) Plahvatav Kõrvemaa. In: Elstrok, H. (comp.) *Põhja-Eesti südamaadel.* Tapa, Tallinn, pp. 391–393.

Mäemets, A. (1977) *Eesti järved.* Valgus, Tallinn.

Ministry of the Environment (1998) *Guidelines for Development of Management plans for Protected Areas.* Ministry of the Environment, Tallinn.

Reimann, M. (1996) Põhja-Kõrvemaa looduskaitseala. Sotsioloogiline uurimus. Manuscript at the Tallinn Pedagogical University.

Reimann, M. (1998) Põhja-Kõrvemaa maastikukaitseala ja selle rekreatiivsed võimalused. Manuscript at the Tallinn Pedagogical University.

Tõnisson, A. (1984) Üks vana piirikivi. *Eesti Loodus* 2, 99–107.

Tõnisson, A. (1991) Põhja-Kõrvemaa looduslikud tingimused ja funktsionaalne tsoneerimine. Manuscript at the Tallinn Botanical Gardens.

4

Hypotheses about Recreational Congestion: Tests in the Forest of Dean (England) and Wider Management Implications

Colin Price and T.W. Mark Chambers

This contribution is substantially based on a paper first printed in *Journal of Rural Studies*, Vol. 2: T.W. Mark Chambers and Colin Price, 'Recreational congestion: some hypotheses tested in the Forest of Dean', pp. 41–52, copyright 1986. The material is reproduced with permission from Elsevier Science.

Introduction

In the era of multipurpose forestry, providing recreational opportunities rightly takes an important place among objectives for forest management. Such an outcome was presaged in the 1972 forestry cost–benefit study (Treasury, 1972): this projected 10% per year growth for visits to Forestry Commission forests, from a then-estimated 15 million visits annually, and suggested that recreation would soon provide greater social benefit than timber production. Since then, recreational provision has been used as a major argument for promoting forest expansion, particularly in the form of community forests (Countryside Commission, 1987).

However, it is worth asking what recreational advantage accrues from designating more facilities, redesigning forests, and even expanding forest area. In some cases – particularly the community forests – a new recreational location is being offered, close to populations. But, where forests and facilities exist already, what is the need to expand them?

One answer is to provide a different type of facility for an increasingly sophisticated and stimulus-seeking clientele. But another purpose of expansion may be that existing forests and facilities have reached, at least during busy periods, their carrying capacity: maintaining the existing quantity and quality of experiences then depends on extra provision.

Most recreational use is concentrated at focal points in relatively few forest blocks. Problems of heavy use arise through disturbance of wildlife, trampling of vegetation, erosion of soil and physical limits of such facilities as car parks, toilets and litter bins. These adverse effects are clearly demonstrable. The deleterious psychological effect of crowding is less easily shown. Yet in ecologically robust areas this is the most important reason for dispersing recreational facilities. To justify heavy expenditure on new and dispersed facilities, evidence is required that visitors are actually crowd-averse.

A belief may be discerned that the importance of crowding can be decided by personal impression or by simplistic survey. Such is not the case: crowding is a complex phenomenon, differently perceived by different individuals. The issue – if it *is* an issue – deserves a proper management response: define the problem; investigate it systematically; implement an appropriate management regime. That is what this chapter attempts to address.

Crowding: a Problem or Not?

Forest managers have differed markedly in views on crowding. Some ('the paternalists') assume that recreationists share (or can be induced to share) their own taste for relative solitude, and try, by establishing trails, way-marking and interpretation centres, to encourage public access through the forest. Others ('the elitists') adopt a viewpoint characterized as 'the average motorist does not, thank goodness, stray more than 50 yards from his car'; they favour high intensity recreation areas around tourist honeypots, keeping the woodland interior for 'those able to find their own way by map and compass'.

Yet a third group ('the market researchers') seek to establish what visitors themselves want, by questionnaire surveys (Wagar, 1964; Stankey, 1972; Cicchetti and Smith, 1976; Price, 1979). Their conclusion has been that recreationists are averse to congestion and will pay a premium to avoid it.

However, during the 1970s anomalous results arose from US surveys. When visitors were asked to evaluate their experience under different degrees of crowding, no clear relationship emerged between stated satisfaction with the visit and degree of crowding (Haas and Nielsen, 1974; Shelby and Nielsen, 1975; Manning and Ciali, 1980; Shelby, 1980). Moreover, Vaux and Williams (1977), when evaluating recreation from travel cost data, found no greater imputed value under quiet than under busy conditions.

Even more puzzling were results from a heavily used site (Tarn Hows) in the English Lake District (Brotherton *et al.*, 1977). When visitors evaluated the crowdedness of a site, an inverse relationship appeared between perceived crowding and actual density, at least until density became very high. The more crowded a site, the less the visitors noticed the crowding! Trakolis (1979) also found that perceived crowding did not increase with increasing density until a high threshold was reached. These results apparently discredit the conventional congestion model, and support the elitists: research apparently vindicates the sentiment uncritically verbalized as 'visitors obviously enjoy crowding; there are always more of them at crowded sites'. There is thus no justification for expanding forests and dispersing facilities to alleviate a non-problem.

However, some evidence for the inverse satisfaction–density relationship has been found, particularly at high densities (Andereck and Becker, 1993a) or in perceived wilderness areas (e.g. Herrick and MacDonald, 1992). A positive willingness to pay for low crowding levels appears in contingent valuations (Rollins *et al.*, 1995; Michael and Reiling, 1997) and stated preferences (Morton *et al.*, 1995). Clearly the satisfaction–density relationship is complex, and in recent models social norms and personal expectations are as important as actual density in determining perceived crowding and satisfaction (Whittaker and Shelby, 1988; Vaske *et al.*, 1996; Tarrant *et al.*, 1997). Satisfaction may be reduced by the behaviour of other recreationists (West, 1982), and evidence of their presence – litter, noise, over-use (Anderson and Brown, 1984) – more than by crowding *per se*.

Explanatory Hypotheses

Still, the contradictory results remain. Many years ago, six explanatory hypotheses were advanced (Burton, 1973; Heberlein and Shelby, 1977) to account for them. Over the succeeding time, a certain amount of testing has been attempted, and the lack of controversy suggests that they have been provisionally accepted as 'reasonable', but no large-scale, systematic attempt to confirm or refute them seems to have been made.

Environmental confounding

If research encompasses several sites, or different places within one site, attractive and accessible locations will be more crowded. Yet greater attractiveness and accessibility promote greater satisfaction, outweighing the negative crowding effect, and in single-site surveys, both greater crowding and greater satisfaction are likely on days of good weather. Such confounding factors may be removed by asking individuals to state preferences between images of sites having different crowding levels but otherwise identical. Manning *et al.* (1996) by this means showed a strong relationship between crowding level and subjective rating of the site.

Dissonance

Users choose activities that agree with their idea of a good time. This, together with the fact that people have voluntarily chosen their activity and have invested time and money in it, leads to a positive evaluation of the experience: expressing dissatisfaction with a chosen activity confesses incompetence in making the choice. Thus high satisfaction levels are reported regardless of conditions. Such cognitive responses to crowding (revising norms in the light of experience) have indeed been demonstrated (e.g. Kuentzel and Heberlein, 1992). These findings show that dissonance does induce adaptation to situations where crowding would otherwise cause dissatisfaction. No relationship of the dissonance strategy to the degree of investment seems to have been investigated.

Product shift

At high crowding levels visitors redefine the experience desired to be compatible with existing congestion, e.g. intention to undertake a quiet nature walk may be replaced by participation in a woodland fun-run, if it transpires that one has been organized. This hypothesis has been 'generally supported' by data from studies of water-based recreation: users did change their definitions of desired experiences, towards those compatible with higher density (Shelby *et al.*, 1988; Shindler and Shelby, 1995). Product shift did not completely alleviate adverse response to crowding, however.

No expectations

First-time visitors have no norms for what to expect of site conditions such as crowding, and so are satisfied with whatever they find. That perception of crowding is a function of norms and expectations has become bedrock in recreation sociology (Andereck and Becker, 1993b; Michael and Reiling, 1997). Heberlein (1992) found that hunters given realistic information about what to expect showed less dissatisfaction with crowding. Note, however, that this result does not confirm the original hypothesis, which states that *no expectations*, rather than *realistic expectations*, prompt satisfaction with the recreation experience.

Displacement

Some visitors are particularly crowd-averse, and avoid congested times (Manning and Ciali, 1980) and sites (Becker, 1981); within sites the crowd-averse move furthest from access points (Burton, 1973). Thus recreationists interviewed at peak times and popular locations tend to be crowd-tolerant: they perceive crowding less stringently than do off-peak users, and their satisfaction is less sensitive to the presence of other people.

Differences in response to crowding have certainly been found between on- and off-peak recreation (Michael and Reiling, 1997). Social surveys have recorded active displacement processes associated with crowding (Anderson and Brown, 1984); lower satisfaction with the previous experience exists among those displaced (Robertson and Regula, 1994). As well as controlling for confounding factors, photographic choice experiments (Price, 1979; Manning *et al.*, 1996) avoid displacement biases: a defined set of respondents gives views on sites in different states of crowding. This hypothesis provides a particularly powerful explanation of the most anomalous results, such as the inverse relationship between actual and perceived density at Tarn Hows.

Vegetational and topographic influences

Vegetative cover and moulding of land-form affect the proportion of visitors that can actually be seen (Burton, 1973). Thus perceived crowding is not necessarily greatest at sites of most intense usage. Particularly, it is asserted that forests have special ability to absorb crowds (Bell, 1998), but the literature seems singularly bereft of recent survey work on this important topic.

Overview

Overall, evidence for the hypotheses has grown, giving ample reason to believe that crowding does reduce satisfaction, particularly above thresholds (even though those *are* socially constructed norms). But, disappointingly, research has not been continued in the UK. Published results emanate largely from the USA, where, disappointingly also, little account has been taken of earlier UK work. The research there concentrates on water-based and wilderness recreation, for which norms and thresholds may not be relevant to UK forests.

It may therefore be timely to re-present the results of a study undertaken in the Forest of Dean in 1984 (Chambers and Price, 1986), which a search of literature suggests remains the only systematic UK investigation of the hypotheses. Perhaps, now that recreational provision is firmly established as a major purpose of forestry, further research in this area may be stimulated, and the implications for the current situation may be pondered. The following sections are an edited version of the original paper.

Crowding in the Forest of Dean

With its pleasing variety of topography, tree species and age-classes, and ready accessibility for a large population in South Wales and West England, the Forest of Dean is a focus of recreational activity, attracting early survey work (Colenutt and Sidaway, 1973). Three surveys were carried out in summer 1984 to collect direct evidence on several factors related to crowding (Chambers, 1984): arrival and departure times were logged at five well-used car parks; visitor densities in areas of different vegetative cover were recorded; questionnaires were administered to 491 visitors at four popular sites in the Forest.

The questionnaire, as well as recording perceived satisfaction and various measures of crowding, elicited information on origin and mode of journey, reasons for choosing the destination, expectations of and response to crowding, and response to environmental detractors.

An attempt was made to reproduce previous surveys' results for the satisfaction–density relationship. Crowding was measured in three ways whenever a questionnaire was administered.

- Visitors were asked to classify the site as: 'Deserted / Fairly deserted / Not busy / Comfortable / Busy / Crowded / Packed'. This indicated perceived density.
- Visible density was measured by counting the visitors in sight.
- The problem of determining actual density (people per hectare) lies not in counting visitors, but in determining the boundaries of a hectare in irregular terrain. Therefore, an angle gauge was constructed, defining an angle, θ, subtended by a person of average height standing on the edge of a circle enclosing 1 ha (see Fig. 4.1). A person subtending a greater angle is within the area, and the number of such people counted in a 360° sweep measures actual density. Similar devices can be constructed to measure the numbers in the nearest 0.5 ha, 2 ha or any other desired area. Because the device counts people in the immediate vicinity, it is less susceptible than 'numbers in sight' to screening by vegetation or topography. This has potential in congestion research: a fuller description appears in Price (1971) and Chambers (1984).

Satisfaction with the visit was expressed by visitors on a descriptive scale, 'Very poor / Poor / Ordinary / Fairly good / Very good / Excellent / Perfect'.

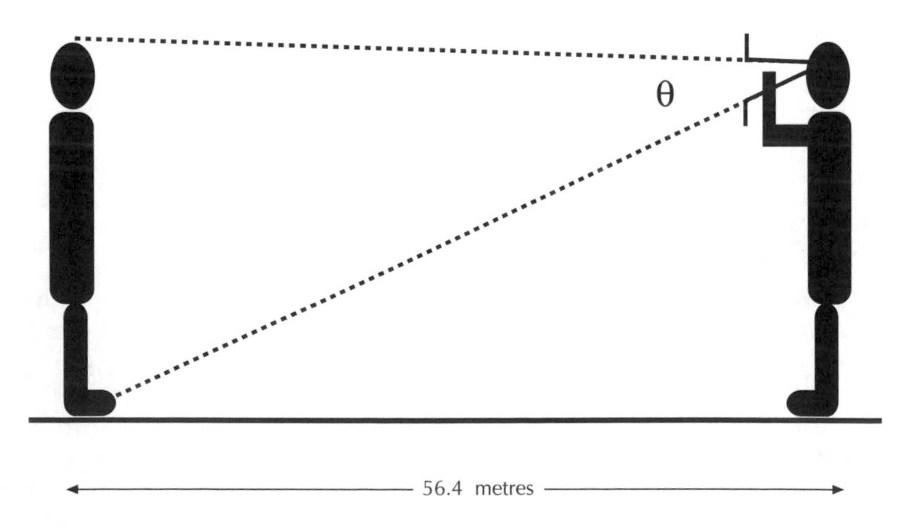

Fig. 4.1. Defining the limits of a hectare using the angle gauge.

Table 4.1. Interrelationships of satisfaction and various measures of density.

Regression of	On	Equation	R^2	Probability
Numbers in sight	√ Actual density	$N = 4.27 + 56 \times \sqrt{D}$	0.534	0.0001
Perceived density	Numbers in sight	$P = 2.96 + 0.00236 \times N$	0.233	0.0001
Satisfaction	Perceived density	$S = 6.39 - 0.0431 \times P$	0.147	0.0001
Perceived density	√ Actual density	$P = 2.95 + 0.195 \times \sqrt{D}$	0.186	0.0001
Satisfaction	Numbers in sight	$S = 5.15 - 0.00972 \times N$	0.020	0.002
Satisfaction	√ Actual density	$S = 5.14 - 0.0541 \times \sqrt{D}$	0.013	0.02

'Probability' means the probability that such a relationship arose by chance.

Relationships between satisfaction and crowding measures were determined by linear regression. Square root of actual density was also used as a variable. Table 4.1 presents regression results for all 491 questionnaires: the regression equation, the R^2 value (the percentage of variance of the dependent variable attributable to the independent variable) and the probability that the relationship is merely random.

A statistically significant inverse relationship exists between satisfaction and each measure of crowding. However, actual density has little impact, accounting for only 1.3% of variance in satisfaction. The first cause of low R^2 values is the indirect and complex relationship between actual density and satisfaction: Fig. 4.2 illustrates some relevant factors. Table 4.1 shows strong relationships between adjacent factors in the causal chain from actual density to numbers in sight to perceived density to satisfaction. However, at each link more variability is introduced, some due to identifiable factors, some unexplained and residual. This weakens the relationship between the ends of the chain.

One approach to investigating such complexity is multiple regression analysis, in which satisfaction depends on many possible explanatory variables, as in Table 4.2. Only perceived density appeared significant among

Table 4.2. Multiple regression of satisfaction on several variables.

Independent variable	Coefficient	F-value	Probability
Perceived density	−0.396	67.3	0.0001
Numbers in sight	−0.00324	0.8	NS
Actual density	−0.000863	0.3	NS
'Wenchford'	0.652	17.0	0.0001
'Mallards Pike'	0.422	9.8	0.002
'Beechenhurst'	0.151	1.4	NS
Distance of origin	0.121	7.8	0.006
Off-peak/peak	0.0511	0.3	NS
Weekend/weekday	−0.414	16.3	0.0002
Group size	0.0240	7.5	0.01
Expectation of crowds	−0.0978	1.1	NS
Constant	6.55	377.6	0.0001

NS = relationship not statistically significant.

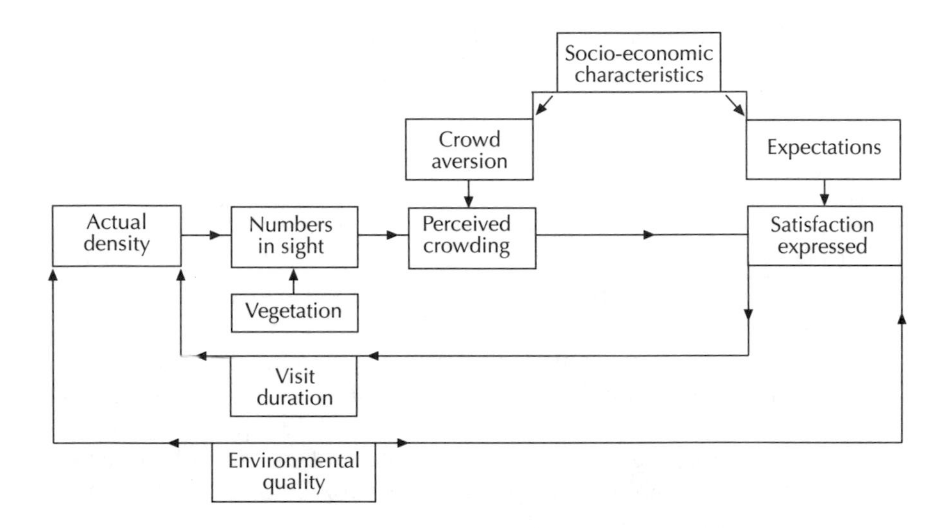

Fig. 4.2. Interaction of variables affecting density and satisfaction: arrows show the direction of hypothesized influence.

the measures of density: the other measures signify only as determinants of perceived density. Other significant differences were between Wenchford (a site in broadleaved woodland) and Mallards Pike (in conifer woodland), contrasted with Beechenhurst (a grassland site) and Speech House (also in broadleaved woodland, but at a major crossroads in the Forest). Visitors from greater distances, those in larger groups and weekend visitors also expressed significantly greater satisfaction.

Assumptions of additivity and independence underlie multiple regression analysis: each explanatory variable should exert an effect on the dependent variable (in this case, satisfaction) which is unrelated to that of other explanatory variables. Clearly, however, crowding variables are strongly related. Furthermore, as demonstrated later, the satisfaction and density relationships appear quite different, depending on values taken by other explanatory variables. Multiple regression analysis can, therefore, only be an exploratory tool. A better analytical strategy is disaggregation: that is, dividing the data according to the value taken by one variable, and determining separate relationships for each sub-group of data. This strategy is pursued subsequently. Chambers and Price (1986) discuss it further.

Testing the environmental confounding hypothesis

According to the structure of causality illustrated in Fig. 4.2, environmental confounding acts by influencing density and satisfaction in the same direction. In fact the exceptional summer experienced by south-west England in 1984 meant that none of the data were affected by adverse weather.

Table 4.3. Satisfaction–density regressions disaggregated by site.

Sites in regression	Equation	Sample	R^2	Probability
All	$S = 5.14 - 0.0541 \times \sqrt{D}$	491	0.013	0.02
Speech House	$S = 4.61 + 0.0143 \times \sqrt{D}$	128	0.001	NS
Numbers in sight	$S = 4.63 - 0.0013 \times N$		0.020	NS
Perceived density	$S = 5.08 - 0.107 \times P$		0.011	NS
Wenchford	$S = 5.23 - 0.0437 \times \sqrt{D}$	87	0.015	NS
Mallards Pike	$S = 5.82 - 0.208 \times \sqrt{D}$	125	0.057	0.01
Beechenhurst	$S = 5.36 - 0.137 \times \sqrt{D}$	151	0.051	0.01
All but Speech House	$S = 5.37 - 0.0901 \times \sqrt{D}$	363	0.035	0.0005

However, the four survey sites had significantly different levels of satisfaction and density. To remove environmental confounding, satisfaction–density relationships were re-analysed for sites individually (Table 4.3).

Three sites displayed similar relationships to the combined data. Speech House, however, was very different, no measure of density significantly affecting satisfaction. This might reflect the existence of special facilities, e.g. barbecue pits, which induced a more communal and crowd-tolerant atmosphere. Removing this site improved the significance of all relationships.

Differences in attractiveness between sites did not induce a systematic positive relationship between density and satisfaction – the site where mean satisfaction was greatest was least crowded, the second best, most crowded. However, sites differed in the kind of visitor attracted and response to crowding: particularly, the rate of decline in satisfaction with density, indicated by the slope coefficient of density variables, differs greatly. In the aggregated data, this large residual variation obscures the underlying relationships.

Testing the vegetational influence hypothesis

Vegetation affects the relationship between actual density and numbers seen, and hence perceived density. Table 4.4 gives regressions for individual sites, which presented different vegetational conditions. At each site a significant relationship existed between actual density and numbers seen, and, at all but one, between perceived and actual density. The statistical strength of relationships, however, was much less at broadleaved sites (large boles and some understorey) than at either conifer (high canopy, little understorey) or grassland (little barrier to visibility) sites.

On Sundays between 14.00 and 16.00 hours, during peak crowding conditions, an experiment was performed to test the vegetational influence hypothesis objectively. At each sample point, one of six angle gauges was selected randomly, the six measuring visitor numbers in the nearest 1/16, 1/8, 1/4, 1/2, 1 and 2 ha. Although measuring density accurately, angle gauges are increasingly susceptible to underestimation in dense vegetation as the area enclosed by the sweep increases.

Table 4.4. Regressions of numbers in sight and perceived density on actual density in different vegetation types.

Site character	Equation	Sample	R^2	Probability
Broadleaved I	$N = 19.0 + 0.224 \times D$	128	0.316	0.0001
Broadleaved II	$N = 32.3 + 0.176 \times D$	87	0.180	0.0001
Coniferous	$N = 7.69 + 0.889 \times D$	125	0.475	0.0001
Grassland	$N = 14.7 + 0.604 \times D$	151	0.491	0.0001
Broadleaved I	$P = 3.74 + 0.00650 \times D$	128	0.069	0.005
Broadleaved II	$P = 3.98 + 0.00203 \times D$	87	0.015	NS
Coniferous	$P = 2.68 + 0.0648 \times D$	125	0.239	0.0001
Grassland	$P = 3.25 + 0.0231 \times D$	151	0.170	0.0001

The regression of observed density on area enclosed was not significant, owing to the extreme variability of the data obscuring any systematic relationship. (Visitors tend to cluster in family or activity groups. Measurement close to these groups yields very high densities, especially when a small area is enclosed by the sweep: readings outside these structural groups give very low densities.) Nonetheless, broadleaved areas seemed rather more successful in absorbing recreationists.

The difficulty could be overcome by measuring density with a gauge defining, simultaneously, limits to two different ground areas. The ratio between the two counts would provide data less susceptible to random variation.

Testing the displacement hypothesis

Displacement of crowd-averse by crowd-tolerant visitors complicates the satisfaction–density relationship, because respondents at times and places of high density usage have different socio-economic characteristics and preferences from those responding under low densities. Burton (1973) demonstrated differences in socio-economic characteristics between peak and off-peak visitors to Cannock Chase. Our data were gathered to quantify how far displacement might bias the satisfaction–density relationship.

Firstly, crowd-averse visitors may select a time of uncrowded conditions. Of the 218 respondents who gave a definite reason for visiting at that particular time, 143 said they were looking for quietness. Moreover, the 72 of these explicitly crowd-averse visitors who came off-peak (early morning and evening) was very significantly higher than the proportion of other visitors who came off-peak (101 out of 348).

Table 4.5. Regressions of satisfaction on perceived density: off-peak and peak.

Time of day	Equation	Sample	R^2	Probability
Off-peak	$S = 6.56 - 0.45 \times P$	177	0.21	0.0001
Peak	$S = 6.25 - 0.35 \times P$	314	0.11	0.0001

Table 4.6. Regressions of stated stay-time, *T*, on various density measures.

Regression on	Equation	R^2	Probability
Perceived density	$T = 138 + 7.89 \times P$	0.7	NS
Numbers in sight	$T = 137 + 1.23 \times N$	5.0	0.0001
Actual density	$T = 159 + 0.40 \times D$	2.0	0.002

Further evidence of displacement comes from regressions of satisfaction on perceived density for peak and off-peak visitors (Table 4.5). While both relationships are highly significant, the satisfaction of off-peak users displays greater sensitivity to perceived density, and perceived density explains almost twice as much variance in satisfaction.

'A user who arrives at a particular location to find it unacceptably congested may leave for some other, perhaps less crowded location' (Burton, 1973). Eleven separate surveys of car arrival and departure times were made to test this at five car parks. Movements of over 2000 cars were logged. In each survey, number of cars already in the park at arrival time had some negative effect on stay-time. The influence was small, R^2 values ranging from 0.008 to 0.096. However, in only one survey was the result not significant at the 0.05 level, while six were significant at the 0.01 level. By contrast, time of arrival had no significant impact on stay-time. Displacement in this sense is real, though not dominant.

By contrast, questionnaire data showed a positive relationship between visitors' estimated stay-time and each measure of density (Table 4.6). This seems to refute the displacement hypothesis. An alternative explanation may be offered, however: on days when visitors decide to stay longer at a site, there are more visitors at the site *at any one time*. If so, actual density, being directly affected by stay-time, would be more strongly correlated with it than is perceived density. On the other hand, if density influences stay-time, actual density affects visitors only indirectly via perceived density, so that the actual density–stay-time correlation should be weaker than the perceived density–stay-time correlation. None of the relationships is strong, as Table 4.6 shows. However, they suggest that stay-time directly influences actual density rather than that perceived density influences (positively) stay-time.

Another form of displacement of crowd-averse by crowd-tolerant visitors is relocation of visits when a site is expected to be crowded: this cannot be observed at the site itself, so no direct research was undertaken. Nevertheless, responses to questions concerning hypothetical changes of crowding indicate the strength of potential displacement. There were 58 respondents who found the site crowded and expected it to be so; they are clearly crowd-tolerant. Of the remaining 433 respondents who answered the question, 255 would have visited another site had they expected crowds, while 18 would have stayed at home. Note particularly the following categories of respondent.

Table 4.7. Satisfaction–density regressions disaggregated by crowd-aversion.

Crowd aversion index	Equation	Sample	R^2	Probability
< 0.8	$S = 5.50 - 0.0068 \times N$	115	0.021	0.1261
0.8 to 1.6	$S = 5.35 - 0.0122 \times N$	147	0.042	0.0129
1.6 to 3.2	$S = 5.66 - 0.0296 \times N$	114	0.173	0.0001
> 3.2	$S = 5.24 - 0.0492 \times N$	115	0.150	0.0001

- *Satisfied with uncrowded conditions, which they expected*
  146 respondents, of whom 66% would not have stayed had the site been perceived as crowded.
- *Satisfied with uncrowded conditions, but had no expectations*
  145 respondents, of whom 56% would not have come if the site was known to be crowded.
- *Unsatisfied with crowded conditions, which they did not expect*
  39 respondents, of whom 69% would not have come had existing site conditions been known.
- *Satisfied with crowds but did not expect them*
  17 respondents, of whom 88% would have come regardless of crowding conditions.

Each class of user responded consistently with the displacement hypothesis.

Some index of crowd aversion would clearly help to disentangle the impact of displacement. Crowd aversion is evidenced by a high perceived density for a given actual density. A crowd aversion index, CAI, defined as

$$\text{CAI} = 2.5^{[\text{perceived density}]} / [\text{actual density}]$$

yielded the best curve-fits as indexed by R^2.

CAI so defined showed the expected, and statistically significant, relationships with visitors' stated response to crowding. The higher the CAI, the greater the probability that the visitor would have visited another site instead, were the site investigated known to be crowded. The higher the CAI, the smaller the stated increase in crowding required to induce the visitor to move. The satisfaction–density model benefited from including CAI. The data were disaggregated into four crowd-aversion classes, each with approximately equal numbers of visitors. Table 4.7 shows the relationship strengthening with increasing crowd-averseness, and impact of numbers in sight on satisfaction becoming more deleterious. Clearly some visitors find crowding much more important than do visitors in general.

Displacement is itself a negative behavioural response to crowding. The relationships adduced above are, therefore, further direct evidence for a negative satisfaction–density relationship, as well as explaining the weakness of satisfaction–density correlations in previous studies.

Testing the no-expectations hypothesis

Absence of expectations might make visitors satisfied with whatever crowding they find, thus suppressing a potentially negative relationship. Our survey found no significant difference in mean satisfaction between those who did and those who did not expect the crowding they found, nor was expectation significant in multiple regression. One explanation is that visitors who do expect the levels of crowding they find have made trip decisions on the basis of expectations which subsequently are vindicated: they too may be satisfied with what they find.

However, when the data were disaggregated, striking differences emerged between those with and those without expectations (Table 4.8). The regression of S on N shows no evidence at all that those without expectations are adversely affected by seeing other visitors. This is despite the fact that they clearly perceive a greater degree of crowding (regression of P on N) and are adversely affected by their perception of crowds (regression of S on P). By contrast, those who expected the conditions they found display a significant negative response on S to N as shown by high R^2 and low probability. The decline of satisfaction with increased numbers in sight is ten times that of non-expecters.

The least satisfied visitors should be those who had expectations, but whose expectations were disappointed. The survey did not separate them from visitors with no expectations at all. This indicates a refinement for future surveys.

Overall, the results show that visitors who expect the conditions they find, while achieving similar satisfaction to those who do not, are far more sensitive to crowding. This represents good evidence for the no-expectations hypothesis.

Testing the dissonance hypothesis

Table 4.9 shows significant positive correlation between satisfaction and distance travelled. This may demonstrate just a 'screening' effect, only those who anticipate high satisfaction being prepared to travel long distances. Even

Table 4.8. Satisfaction regressions and the no-expectations hypothesis.

Category of visitor	Equation	Sample	R^2	Probability
Expecting the conditions found	$S = 5.35 - 0.0145 \times N$	250	0.062	0.0002
	$S = 6.51 - 0.415 \times P$		0.175	0.0001
	$P = 2.86 + 0.0321 \times N$		0.297	0.0001
Not expecting the conditions found	$S = 4.91 - 0.00145 \times N$	241	0.001	NS
	$S = 6.24 - 0.371 \times P$		0.119	0.0001
	$P = 3.08 + 0.0245 \times N$		0.165	0.0001

Table 4.9. Regressions of satisfaction and density, disaggregated by distance of origin.

Distance zone (km)	Equation	Sample	R^2	Probability
0–20	$S = 7.23 - 0.56 \times P$	162	0.236	0.0001
20–60	$S = 6.59 - 0.33 \times P$	194	0.136	0.0001
60–180	$S = 4.97 - 0.23 \times P$	71	0.030	0.1362
> 180	$S = 6.39 - 0.31 \times P$	64	0.102	0.0112
0–20	$P = 2.11 + 0.026 \times N$	162	0.245	0.0001
20–60	$P = 4.34 + 0.020 \times N$	194	0.200	0.0001
60–180	$P = 3.20 - 0.036 \times N$	71	0.200	0.0047
> 180	$P = 3.75 + 0.036 \times N$	64	0.300	0.0004

so, those who travel furthest have invested most in the visit. Therefore, according to the dissonance hypothesis, they should be least inclined to express dissatisfaction with crowding. Table 4.9 presents disaggregated regressions.

Visitors from distant origins perceive crowding at least as acutely as those from close at hand. However, the most significant relationships between satisfaction and perceived density occur for the closer origins. Taken together, these relationships support the dissonance hypothesis.

Crowding in relation to other detractors

Respondents were asked 'Which, if any, of the following have detracted from your visit: presence of sheep; aircraft noise; too many other visitors; insects; presence of conifers; too few facilities; presence of dogs, horses, etc.; litter; noticeable vandalism?' These detractors were the most frequently mentioned in a previous survey (Trakolis, 1979). 'Too many other visitors' caused some concern to more visitors (66%) than any other, while litter – a by-product of intensive use – was the second commonest cause (53%). By contrast. 'too few facilities' caused concern to only 9% of visitors. Moreover, when asked to rate their concern, visitors gave crowding the highest rating. Questions on crowding earlier in the questionnaire may have sensitized respondents to this particular detractor, so the responses should be treated cautiously. This question could have been placed earlier in the questionnaire – where, however, it would hint that crowding *should* cause dissatisfaction.

Conclusions from the Survey

While the significance of individual tests may be doubted, the cumulative evidence is very strong that the satisfaction of many summer visitors to the Forest of Dean was adversely affected by crowding. It was shown by their behaviour, their stated attitudes to existing crowding and their responses to hypothesized increased crowding. Furthermore, it proved possible to identify certain groups – those at Mallards Pike and Beechenhurst, off-peak visitors, those with expectations about site congestion, and visitors from local origins – for whom crowding caused serious dissatisfaction, although others were indifferent. Recreation managers need to recognize the preferences of these different groups.

The contrast with results which found no satisfaction–density relationship can be explained variously. Firstly, as compared with a once-in-a-lifetime visit to, say, the Grand Canyon (Shelby, 1980), the Forest of Dean offers, scenically, a relatively common experience for most visitors: many non-local as well as local people have visited the Forest before, and specifically seek an informal and peaceful environment. Many visitors have expectations about crowding levels, and such visitors are crowd-sensitive: the no-expectations hypothesis, which this study confirms, may operate more powerfully elsewhere.

Secondly, the uniformly fine weather removed a potential cause for a positive relationship between satisfaction and intensity of usage, while all the sites investigated were relatively attractive and accessible for recreation: environmental confounding was minimized. Thirdly, for most visitors the trip cost relatively little, and, if it proved unsatisfactory, they would say so: dissonance does not operate strongly. Fourthly, a non-evaluative scale of density was used (deserted–packed), cf. Trakolis's scale (too few–too many) so visitors would not feel that high perceived density was inconsistent with their decision to participate.

Finally, 'until density approaches the level which is interpreted as crowded, it is likely that density causes no social dysfunction' (Manning and Ciali, 1980). This seems relevant, especially as many other studies either admit that insufficient numbers of visitors were recorded (Haas and Nielsen, 1974; Trakolis, 1979; Shelby, 1980) or were carried out in low-density wilderness settings. Sites in the Forest of Dean had clearly reached the threshold where crowding was perceived, and, interestingly, the least significant results from car-park surveys occurred on days of least use.

The above results underline the complexity of the satisfaction–density relationships, while demonstrating that it is possible to separate and quantify some of its components. They also show that the situation in UK forests does not mirror very precisely that investigated by US research, whose results cannot therefore be imported uncritically. The time seems ripe for further UK investigations, in different types of forest.

Management Implications

First, recreation planners may find encouragement in the general attitude of forest visitors to this survey. Of 495 visitors approached, only four refused to answer the questionnaire. Most seemed genuinely pleased that their views were being solicited. Many posed questions about forestry after the formal interview. This emphasizes the value of simple interpretation displays in conjunction with forest recreation facilities. The British public are clearly interested in less superficial aspects of forestry than mere provision of car parking places.

Second, most literature on recreational crowding has as its management end-point the setting of use limits for sites. However, there are also land-use implications. After all, if congestion is not a significant problem, very small or linear forests suffice for recreation purposes, and the recreational case for forest expansion vanishes. It may be worth identifying different types of forest, to which the conclusions on crowding apply with varied force.

High intensity recreation sites

Since the 1980s, community forests close to urban centres have replaced remote upland forests as the focus of national forest recreation strategy. Although the Forest of Dean is more heavily wooded than the community forests (only 5% of whose project area is presently under trees), it resembles them in closeness to major population centres, intensity of use, mixture of species and the importance of open areas: lessons could be transplanted.

- The results of investigating vegetational influence imply that broadleaved cover, with a sufficiently light canopy to allow a substantial understorey, is the best way to absorb crowds. (In fact, broadleaved cover is already prevalent in such forests, but early silvicultural intervention to encourage understorey development is important.)
- At often-visited sites, 'realistic expectations' can replace 'no expectations' as a cognitive basis for reasonable satisfaction. As these forests are a local resource, there is every prospect of such expectations being formed. Plainly, however, the resource should not be oversold as a means of achieving wilderness-type experiences.
- The Dean study clearly revealed the existence of crowd-averse sub-populations. Behavioural displacement to times and locales of low-intensity use allows such incorrigibly crowd-averse visitors to meet their needs. The tendency of different sub-groups to segregate into areas of different use intensity should not be thwarted by spreading recreational pressures and the facilities that attract them evenly throughout the forest, or over time. On the contrary, it is a sound management strategy to 'seed' displacement processes, by providing accessible and attractive areas where the crowd-tolerant can congregate (cf. Speech House), leaving the less 'improved' hinterland for the crowd-averse. Whatever the intuitions of economists, nothing more is required to regulate visitor distribution within a site: because of the costs of regulation, free access may be better than regulation, even under crowding (Price, 1981).

Multipurpose forests

Meanwhile, recreation remains a potential justification of forestry within its new multipurpose ethos. Some large conifer forests of the 20th century show localized high-intensity use. 'Localized' is probably how it should remain, with perhaps three zones being recognized.

- Close to access points are the toilets, cafés, play facilities and short walks. For those visitors who prefer not to stray far from their cars, a continued expansion of facilities avoids the outrunning of physical capacity, while maintaining the sociable, crowd-tolerant recreation experience demonstrated at the Speech House site.
- Where moderately large concentrations of visitors move out into the forest, high-cost experiences can be provided efficiently. The internationally famous sculpture trail at Grizedale Forest in the English Lake District is an example. Such facilities permit the product-shift strategy, under which a would-be solitude seeker can redefine the trip, and follow the crowds through a cheerful art-*fest*.
- For the more adventurous, extension of low-level facilities such as discretely way-marked trails seems appropriate and adequate. For such visitors in the Dean study, presence of crowds caused more concern than lack of sophisticated facilities. Thus in Grizedale the undeveloped southern part of this forest remains, and should remain, for crowd fugitives.

Commercial forests

As for dominantly commercial forests such as Kielder on the England–Scotland border, they may have their high capacity facilities, as around the margins of Kielder Water Reservoir. Given the diversity of crowd-adverseness, it would be a mistake even to attempt a diffusion through the whole vast area.

Nor should the ability of forests to absorb crowds be relied upon uncritically as a means of improving the countryside's recreational carrying capacity, or as a justification for forest expansion, in such areas. Uniform conifer plantations are impenetrable in early life, so that visitors congregate along roads and in open areas, while in maturity the low light levels at the forest floor inhibit understorey development. But if no facilities are provided, there will be no crowds to hide. The unrelieved and uncrowded expanses of monocultural Sitka spruce and large-scale clear-felled sites then provide a different recreation experience, and advantages arise in keeping them as they are. There is nothing improper, even in a multipurpose era, in regarding them primarily as a timber production enterprise: it offers escape from crowds for those who prefer solitude and self-reliance, to the stimulus of provided facilities and the sociability of shared experience.

Acknowledgements

We thank those who assisted in gathering survey data. We are also grateful to Elsevier Scientific, the publishers of *Journal of Rural Studies* in whose pages the original paper appeared, for permission to reproduce it here.

References

Andereck, K.L. and Becker, R.H. (1993a) The effects of density on perceived crowding in a built recreation environment. *Journal of Applied Recreation Research* 18, 165–179.

Andereck, K.L. and Becker, R.H. (1993b) Perceptions of carry-over crowding in recreation environments. *Leisure Sciences* 15, 25–35.

Anderson, D.H. and Brown, P.J. (1984) The displacement process in recreation. *Journal of Leisure Research* 16, 61–73.

Becker, R.H. (1981) Displacement of recreational users between the Lower St. Croix and Upper Mississippi Rivers. *Journal of Environmental Management* 13, 259–267.

Bell, S. (1988) *Design for Outdoor Recreation.* E. & F. Spon, London.

Brotherton, D.I., Maurice, O., Barrow, G. and Fishwick, A. (1977) *Tarn Hows — an Approach to the Management of a Popular Beauty Spot.* Countryside Commission, Cheltenham.

Burton, R.C.J. (1973) A new approach to perceptual capacity. *Recreation News Supplement* 10, 31–36.

Chambers, T.W.M. (1984) Some effects of congestion on outdoor recreation with special reference to the Forest of Dean. Unpublished MSc thesis, University of Wales, Bangor.

Chambers, T.W.M. and Price, C. (1986) Recreational congestion: some hypotheses tested in the Forest of Dean. *Journal of Rural Studies* 2, 41–52.

Cicchetti, C.J. and Smith, V.K. (1976) *The Costs of Congestion.* Ballinger, Cambridge, Massachusetts.

Colenutt, R.J. and Sidaway, R.M. (1973) Forest of Dean day visitor survey. *Forestry Commission Bulletin* 46.

Countryside Commission (1987) *Forestry in the Countryside.* Countryside Commission, Cheltenham.

Haas, J.E. and Nielsen, J.M. (1974) *A Proposal for Determining Sociological Carrying Capacity of the Grand Canyon–Colorado River Area.* Human Ecology Research Service, Boulder, Colorado.

Heberlein, T.A. (1992) Reducing hunter perception of crowding through information. *Wildlife Society Bulletin* 20, 372–374.

Heberlein, T.A. and Shelby, B. (1977) Carrying capacity, values, and the satisfaction model: a reply to Greist. *Journal of Leisure Research* 9, 142–148.

Herrick, T.A. and McDonald, C.D. (1992) Factors affecting overall satisfaction with a river recreation experience. *Journal of Environmental Management* 16, 243–247.

Kuentzel, W.F. and Heberlein, T.A. (1992) Cognitive and behavioral adaptations to perceived crowding: a panel study of coping and displacement. *Journal of Leisure Research* 24, 377–393.

Manning, R.E. and Ciali, C.P. (1980) Recreation density and user satisfaction: a further exploration of the satisfaction model. *Journal of Leisure Research* 12, 329–345.
Manning, R.E., Lime, D.W., Freimund, W.A. and Pitt, D.G. (1996) Crowding norms at frontcountry sites: a visual approach to setting standards of quality. *Journal of Leisure Sciences* 18, 39–59.
Michael, J.A. and Reiling, S.D. (1997) The role of expectations and heterogeneous preferences for congestion in the valuation of recreation benefits. *Agricultural and Resource Economics Review* 26, 166–173.
Morton, K.M., Adamowicz, W.L. and Boxall, P.C. (1995) Economic effects of environmental quality change on recreational hunting in northwestern Saskatchewan: a contingent behaviour analysis. *Canadian Journal of Forest Research* 25, 912–920.
Price, C. (1971) *Social Benefit from Forestry in the U.K.* Department of Forestry, Oxford University.
Price, C. (1979) Public preference and the management of recreational congestion. *Regional Studies* 13, 125–139.
Price, C. (1981) Charging versus exclusion: choice between recreation management tools. *Environmental Management* 5, 161–175.
Robertson, R.A. and Regula, J.A. (1994) Recreational displacement and overall satisfaction: a study of central Iowa's licensed boaters. *Journal of Leisure Research* 26, 174–181.
Rollins, K., Wistowsky, W. and Jay, M. (1995) Wilderness canoeing in Ontario: using cumulative results to update dichotomous choice contingent valuation offer amounts. Working Paper, Department of Agricultural Economics and Business, University of Guelph, DP95–03.
Shelby, B. (1980) Crowding models for back country recreation. *Land Economics* 56, 43–55.
Shelby, B. and Nielsen, J.M. (1975) *Use Levels and User Satisfaction in the Grand Canyon.* Human Ecology Research Service, Boulder, Colorado.
Shelby, B., Bregenzer, N.S. and Johnson, R. (1988) Displacement and product shift: empirical evidence from Oregon rivers. *Journal of Leisure Research* 20, 274–288.
Shindler, B and Shelby, B. (1995) Product shift in recreation settings – findings and implications from panel research. *Leisure Sciences* 17, 91–107.
Stankey, G.H. (1972) A strategy for the definition and management of wilderness quality. In: Krutilla, J.V. (ed.) *Natural Environments.* Resources for the Future, Washington, DC.
Tarrant, M.A., Cordell, H.K. and Kibler, T.L. (1997) Measuring perceived crowding for high-density river recreation: the effects of situational conditions and personal factors. *Leisure Sciences* 19, 97–112.
Trakolis, D. (1979) The concept of carrying capacity of forest recreation areas. Unpublished PhD thesis, University of Wales, Bangor.
Treasury (1972) *Forestry in Great Britain: an Interdepartmental Cost/Benefit Study.* HMSO, London.
Vaske, J.J., Donnelly, M.P. and Petruzzi, J.P. (1996) Country of origin, encounter norms, and crowding in a frontcountry setting. *Leisure Sciences* 18, 161–176.
Vaux, H.J., Jr and Williams, N.A. (1977) The costs of congestion and wilderness recreation. *Environmental Management* 1, 495–503.

Wagar, J.A. (1964) The carrying capacity of wildlands for recreation. *Forest Science Monograph* 7.
West, P.C. (1982) Effects of user behavior on the perception of crowding in back-country forest recreation. *Forest Science* 28, 95–105.
Whittaker, D. and Shelby, B. (1988) Types of norms for recreation impacts: extending the social norms concept. *Journal of Leisure Research* 20, 261–273.

5

Balancing Tourism and Wilderness Qualities in New Zealand's Native Forests

Geoffrey Kearsley

Introduction

New Zealand's native forests cover around a quarter of the country's land area; they are the remains of an almost total pre-human settlement cover of native species that are botanically unique and that harbour a distinctive wildlife. These native forests are almost entirely protected, in a more or less pristine state, by a system of National Parks and Forest Parks that emphasizes conservation above any other use, including recreation. Indeed, so extensive is the forest cover within this Conservation Estate and so comprehensive is the latter that the two are effectively one and the same thing. Only the highest mountain peaks are unforested. Because production and logging of any kind has largely been eliminated and transferred to extensive commercial forests, mainly of exotic pine and other conifers, the principal management issue within the forests is the management of recreation and the restoration of as much of the original wildlife ecology as possible. This latter is achieved mainly through attempts to control introduced species, such as browsers, in the form of deer, pigs and opossums, or predators of forest birds, such as weasels, rats and ferrets, and to rescue endangered species from extinction through a system of sanctuaries and scientific breeding programmes.

Recreational use of the forests traditionally included hunting for deer and pigs or fishing the numerous lakes and rivers to be found within their confines. Hiking, known as tramping within New Zealand, has always been a major activity, both for its own sake and as a means of reaching the Alpine peaks. More recently, adventure activities such as white water rafting and caving have been added. Since the early 1980s increasing numbers of overseas tourists have come to use the forests, especially for tramping, and their

(eds Xavier Font and John Tribe)

numbers are such that issues of crowding and displacement have become major problems for forest managers. This is especially so because most visitors expect to find a high degree of wilderness in the back-country environment. Physical impacts occur (Ward and Beanland, 1996), but these are minor when compared with social ones (Kearsley *et al.*, 1998), so that the environmental management of New Zealand's forest consists almost entirely of social impact mitigation or wildlife protection. Because the remnants of endangered species are either in off-shore sanctuaries or in the remotest of locations, the two tasks almost never coincide in a spatial sense.

This chapter describes how New Zealand's forests were extensively cleared, both by Maori and European settlers, and how a conservation ethic replaced the production ethos of the early pioneers, leading eventually to the protection of native forests in the Conservation Estate. It then goes on to describe the changing use of native forests and the rise of tourism as a significant user. The perceived impacts of rising recreational and tourist use are analysed through discussion of a major survey of back-country users. The chapter concludes by describing the current methods used to manage recreation and by suggesting a method whereby individual perceptions of wilderness can be used to accommodate a large user population in environments that they themselves regard as wild.

New Zealand's Forests

New Zealand was once a land of forests. Isolated by continental drift for approaching 80 million years, a unique ecology developed, characterized by distinctive trees and ferns and by a fauna without predators, dominated by flightless birds such as the kiwi, kakapo and the now extinct moas. When the first Polynesian settlers arrived, about a thousand years ago, almost all of the country was forested; only the highest mountains and some tracts of native tussock grassland were not covered (McGlone, 1983). Maori burning (Cumberland, 1965) and agricultural clearance had reduced this almost ubiquitous cover by half when formal European settlement began in 1840 (Cameron, 1984). Today, less than a quarter of the indigenous forests remain, largely in the most mountainous and least accessible parts of the country. Much of the lowland podocarp hardwood has gone, but the montane beech forests in the South Island are largely intact. Extensive tracts of exotic pine forest (mainly *Pinus radiata*) have been planted and these have become the focus of the pulp and timber industries. The cutting of native forest has largely ceased (Memon and Wilson, 1993), but virtually uncontrolled browsing by introduced deer and opossums continues unabated, while, as noted above, similarly introduced rats, weasels and stoats continue to decimate native birds.

Forest conservation

The pioneer development ethic of early European settlement saw extensive forest clearance, especially on the best pastoral land; while the Forest Act 1874 espoused the principles of sustainability, cutting continued unhindered until the 1960s (Wilson, 1991). Often, even productive forest was seen as 'waste' land until it could be developed, but difficulties of access and transportation meant that much of the more scenically spectacular 'waste' land was untouched before conservation measures were introduced.

Various agencies have been responsible for overseeing forestry, mostly in the interests of production. The Lands Department of the 19th century led to the State Forest Service of 1921 and the New Zealand Forest Service of 1949. Each was responsible for native and exotic forests alike, and, while responsible for introducing sustainable and regenerative forest policies, the focus of these was primarily on long-term continuous harvesting. The 1970s saw substantial environmental debate in New Zealand. At its peak, the Maruia Declaration 1977 was supported by 340,000 signatories (a tenth of all New Zealanders) and called, effectively, for an end to native forest logging. There were many other campaigns to save specific stretches of forest during this period. In 1987, partly as a result of this pressure, the Government abolished the NZ Forest Service and the Department of Lands and Survey and handed the control of almost all native forests to the newly formed Department of Conservation. The Resource Management Act 1991 has added substantial protection for smaller tracts of native forest on private land. Production (exotic) forests were corporatized or sold to private interests.

Most of the land protected by the Department of Conservation is held in the form of National Parks, Forest Parks and other reserves, which, in total, cover about a third of the country. The National Park movement was stimulated by the establishment of Yellowstone in the early 1870s, and began in 1887 with the gift to the Government of the Tongariro volcanoes by local Maori interests. This triggered a process of legislation and land acquisition, usually under the Land Act 1892 and the Scenery Preservation Act 1903, which led to the gazetting of extensive forest areas as scenic reserves, so that a half of the current land area was listed by 1914. The National Parks Act 1952 formalized this process and introduced careful management, including the notion that Parks were to be protected in a natural state. Conservation was to be the primary purpose of the Parks, with recreation and tourism only allowed where conservation values are not compromised. Today, National Parks continue to be created and existing ones enlarged, with very large areas of substantially unmodified country continuing to be added; the latest National Park, Kahurangi, in the north-west of the South Island, is the country's second largest, and was only opened in 1996. Save for the highest peaks, the Parks are almost entirely forested, and relatively little unmodified forest is outside the Conservation Estate.

Demand for forest recreation

From the earliest days of European settlement, the back-country has been used for tramping, climbing, skiing, hunting and fishing, and, as a consequence, an outdoor recreational ethic has been a substantial component of New Zealand's culture and way of life (Fitzharris and Kearsley, 1987). However, the demand for natural areas where outdoor recreation can take place has increased dramatically in recent years. In New Zealand an increase in international tourism over the past two decades has seen visitor numbers rise from around 250,000 in the mid 1970s to 1.5 million in 1999. This sixfold increase has put pressure on resources that were traditionally utilized by predominantly domestic recreationists and has come to threaten the very resources upon which both recreation and tourism are based, the mountains and forests preserved in the country's National Parks and other reserves. Not only have numbers increased, but, as tourists require a much greater direct experience than was provided by the original scenic tours, the number of visitors entering the forested back-country has increased enormously. In particular, tourist use of the back-country is focused upon a series of walking tracks, the most popular of which are known as the 'Great Walks'.

With this increase in demand has come a range of associated problems. Although physical impacts upon the natural environment can be recognized (Ward and Beanland, 1996; Kearsley and Higham, 1997), crowding is the more serious impact, affecting the very nature of the outdoor recreation experience itself. In recent years, several studies have suggested that the rise of overseas visitors has begun to impact upon the more established and popular parts of the Conservation Estate and to generate perceptions of crowding (Kearsley, 1990; Keogh, 1991; Higham and Kearsley, 1994; Kearsley and O'Neill, 1994; Kearsley, 1996). Kearsley *et al.* (1998) show that up to 54% of visitors experience some degree of crowding, while more than two-thirds of some sub-groups do so.

As a result of this, the suspicion has arisen that some domestic, and possibly some of the more adventurous overseas, trampers are being displaced into marginal environments or seasons so as to avoid perceived crowding. One consequence of this, if it is happening, is increased visitor pressure on more remote locations and displacement of people with only moderate wilderness images (Kliskey and Kearsley, 1993) into a limited reservoir of pristine sites, with obvious physical impacts. Similarly, there will be an impact on host community satisfaction as overseas visitors displace domestic recreationists. Both of these consequences will have implications for the sustainability of tourism in New Zealand.

The nature of wilderness in New Zealand

Much of the motivation for visiting the New Zealand back-country and its forest cover is to experience wilderness (Kearsley and Higham, 1996).

Wilderness can be considered from many perspectives; one is that it is simply an environment and ecology undisturbed by human action. By this definition there would be little wilderness left in New Zealand, as the depredations of deer, goats and opossums attack forests from below and above, and as stoats, cats, dogs and weasels continue to decimate wildlife almost everywhere. The effects of global warming, ozone depletion, residues from nuclear tests and the widespread dispersion of agricultural chemicals and pesticides have altered, and will alter, the balance of unmodified and pristine nature.

Wilderness has a legislative definition that was born with the Wilderness Act 1964 in the United States, which observes that wilderness is 'an area where the earth and its community of life are untrammelled by man, where man himself is the visitor that does not remain'. It specifies that wilderness areas should be affected primarily by the forces of nature, should provide opportunities for solitude and primitive recreation, should be large (at least 5000 acres) and might contain ecological or geological features of scientific, educational, scenic or historical value.

These views were taken up by the National Parks Authority of New Zealand, who defined wilderness as 'an area whose predominant characteristic is the interplay of purely natural processes large enough and so situated as to be unaffected, except in minor ways, by what takes place ... around it'. This definition has been elaborated by others who have cited specific physical criteria that require a wilderness to have, for example, an area of more than 20,000 ha, a diameter of 2 days travel time and a buffer zone of a day's travel around it (Molloy, 1983).

These requirements have been enshrined in the provisions of various pieces of significant legislation, especially the Conservation Act 1987. Nevertheless, almost all writers, including Molloy (1983), recognize that an essential part of the value of wilderness to the individual lies in the emotions and state of mind that are stirred in that person by the wilderness experience. People themselves experience wilderness in many different settings, not simply formally designated Wilderness Areas, while changing cultural attitudes to wilderness have been well documented (Glacken, 1967; Nash, 1982; Oelschlaeger, 1991; Shultis, 1991; Hall, 1992). As will be demonstrated at the end of the chapter, it is this multiplicity of images that makes it possible to manage the wilderness experience for large populations, with lessened risk of irreversible environmental damage.

Impacts of Back-country Forest Recreation

Since the control of the effects of recreation and tourism is the principal concern of native-forest managers, the remainder of this chapter reports some of the results of a questionnaire survey made possible by funding from the New Zealand Foundation for Research, Science and Technology Public Good

Science Fund and carried out during the tramping season of 1995/6. Some 950 back-country users were contacted in the field and invited to take and subsequently complete and return a self-completion mail-back questionnaire. Respondents were contacted throughout the whole of the country and in a wide range of back-country environments, ranging from the highly popular Great Walks, such as the Routeburn, Abel Tasman and Kepler Tracks, to scarcely used wilderness routes. Half of the respondents were New Zealand residents and half were international visitors. The aim of the survey was to measure perceptions of crowding, motivations and degrees of satisfaction with the experiences gained and to gain some sense of the extent to which displacement and coping strategies were taking place in a large sample.

Motivations

The motives for tramping and otherwise using natural environments have been analysed in many past studies (Moore, 1995) and have been found to be largely consistent over time. Those found in this study (Table 5.1) are no different. Visitors came to the back-country above all to find naturalness and scenic beauty and to enjoy the outdoors. Significant numbers came to encounter wilderness, and, while relatively few wished to find total solitude, neither did they wish to meet new people and to make friends. The companionship of one's own group is desirable, but not that of too many others. Clearly, the forest recreation experience is not tolerant of large numbers of other users. As can be seen in Table 5.1, there was little difference between domestic and overseas users in terms of the main motivations, but rather more New Zealanders sought physical challenge and the pursuit of personal goals, whereas international visitors emphasized learning and novelty.

Table 5.1. Motivations for visiting natural areas, percentage regarded as important.

	Domestic	Overseas
Scenic beauty/naturalness	92	97
To enjoy the outdoors	96	95
To encounter wilderness	78	82
To undertake physical exercise	72	59
To get away from life's pressures	72	56
To face the challenge of nature	61	65
To relax with family/friends	64	40
To achieve personal goals	55	42
For a totally new experience	41	52
To learn about NZ plants and animals	34	48
To experience solitude	38	39
To meet new people and make friends	16	18

Table 5.2. The overall extent of crowding, per cent.

	Domestic	Overseas
Not at all crowded	36	24
2	15	17
3	15	15
Slightly crowded	15	17
5	5	9
Moderately crowded	7	11
7	4	4
8	1	2
Extremely crowded	3	1

Degrees of crowding

The overall extent to which crowding was perceived is set out in Table 5.2, using a scale developed by Shelby *et al.* (1989) and used elsewhere in New Zealand (Kearsley and O'Neill, 1994). Although the domestic and international figures appear, at first glance, to be very similar, the fact that visitors tended to be on the more popular tracks requires that further analysis be attempted before firm conclusions can be drawn. Nonetheless, in the back-country in general, while 30% overall felt quite uncrowded, some 16% reported moderate to extreme crowding. The apparently higher perception of the absence of crowding by New Zealanders is a reflection of their concentration in the more remote places, itself a possible indicator of displacement. This displacement process appears to be taking experienced domestic users into more remote and ecologically vulnerable locations and encouraging visitors into environments that many are not experienced or well equipped enough to handle.

The extent to which perceived crowding was said to have affected enjoyment is set out in Table 5.3. Twenty-two per cent of the sample said that crowding had affected their enjoyment, and some two-thirds of those said that it had done so moderately to extremely. Overseas visitors were rather more likely to have been affected than domestic users. Crowding was by far the largest impact reported and thus one of the most critical issues for management.

Table 5.3. The extent to which crowding affected enjoyment, per cent.

	Domestic	Overseas
Not at all	83	74
2	5	7
Moderately	8	14
4	2	5
Extremely	1	–

Table 5.4. Perceived impacts that largely or totally spoilt overall enjoyment, per cent.

	Domestic	Overseas
Noise in huts	11	13
Aircraft noise	10	15
Commercial operations	8	13
Jet boat noise	8	11
Untidy huts	9	7
Track standard too high	9	10
Litter on track	10	8
Track widening	7	7
Excessive track wear	7	5
Bunks unavailable	7	5
Accommodation quality too high	6	5
Boardwalks	5	5
Behaviour of hunters	4	3

When the motives for going to the back-country are examined (Kearsley, 1996; Kearsley and Higham, 1996) it can be seen that about 38% of both domestic and international back-country users as a whole see solitude as an important or very important motive for tramping. When asked if specific sites could be identified as being crowded, nearly half were able to name such places, and, while it is not possible to list or analyse them all here, it is noteworthy that most were in fact overnight huts, usually on the more popular walks. Overall, it seems quite clear that a substantial minority had experienced crowding, both among visitors and domestic users.

Other impacts

Apart from crowding, respondents were asked to consider a range of other impacts and to say how far these had spoilt their overall experience. A range of possible impacts was suggested (Table 5.4) and others were offered by respondents in addition to these. Here, it was obvious that it was social rather than biophysical impact that was most widely experienced.

As can be seen in Table 5.4, noise was the predominant irritant, especially for overseas visitors, who tended to be in the busier locations, and who were especially aware of both aircraft (including helicopter) and boat engine noise, as well as noise disturbance in huts. Both groups were aware of litter and untidiness in huts, while the presence of commercial operations was again most noticed by overseas groups. Smaller numbers noticed wear on tracks and widening through muddy areas, but around 10% felt that track standards were sometimes too high, and some objected to the presence of boardwalks, which are used to protect highly vulnerable areas. Other complaints included the unavailability of bunks in high-use huts and the sometimes unnecessarily high standards of hut accommodation.

Many of the people interviewed sought wilderness and wilderness experiences and, as Kearsley (1990) has shown, they tend to associate them with the National Parks, and hence the native-forest environment. Sixty-nine per cent of the sample expected to encounter wilderness conditions; 73% of overseas visitors and 65% of locals expected to do so. However, while most considered that the track that they were on displayed some degree of wilderness character (Table 5.5), few thought that it was pure or pristine wilderness.

Nevertheless, 71% of the sample said that they had, in fact, encountered wilderness in their trip; 73% of New Zealanders and 69% of visitors claimed to have found wilderness, so that most of those who expected wilderness conditions did in fact find them. Of the minority who did not, most said it was because tracks were too well formed, signed and hardened. This view was held by 56% of the overseas visitors who did not encounter wilderness and 40% of locals, reflecting the fact that New Zealanders tended to be encountered in the more remote environments. Overseas respondents were more likely to be on the more developed popular tracks. Thirty-six per cent of each group cited crowding as detracting from wilderness values and a quarter believed that overnight huts were too comfortable and even luxurious, and this, too was a comment mostly associated with the Great Walks. Significant numbers also mentioned boat and aircraft noise.

Perhaps unsurprisingly, those in the remotest locations found pure wilderness most frequently, but international visitors were the group that most frequently found conditions that approached wilderness, with half grading the degree of wilderness encountered at four on a five-point scale, even though they were in the most popular places. It would seem, therefore, that their expectations of wilderness are somewhat more tolerant of human impact than are those of New Zealand residents.

Displacement

Displacement occurs as the result of dissatisfaction with present or past experiences or expectations of likely future conditions and refers to the unwilling movement out of preferred places or times or to the re-evaluation of actual experiences. Displacement may be spatial, when recreationists move to another site in order to obtain a preferred experience, or it may be seasonal

Table 5.5. The extent to which the track exhibited a wilderness character, per cent.

	Domestic	Overseas
Not at all	4	3
2	11	13
3	31	31
4	43	47
Pure wilderness	12	5

(Nielson and Endo, 1977; Anderson and Brown, 1984). Others may reinterpret the meanings and benefits expected from a site in a process of 'product shift', as when a crowded track is seen as providing social experience in a natural area rather than a wilderness encounter. In spatial displacement, those with a low tolerance for crowding, for example, may be displaced by those with a higher.

In a context where there is a clear hierarchy of sites and tracks, as in southern New Zealand, displacement down the hierarchy is an all-too-likely possibility. Thus, one could argue that the very large increases in overseas users of the most famous tracks, such as the Routeburn, and consequent rationing, has displaced some domestic recreationists (and perhaps some tourists) to second- or third-tier tracks or, indeed, out of tramping altogether. Similarly, their arrival might displace others yet further down the hierarchy or into more dangerous seasons, and there is a clear danger that inexperienced trampers might be forced into wild and remote environments and conditions that are beyond their capacity to manage. And, as noted above, such a process might well breed resentment and visitor dissatisfaction.

There is ample evidence to suggest that a fair degree of displacement is going on, in support of the somewhat anecdotal studies reviewed by Kearsley (1995). A fifth of those interviewed (17% of visitors and 23% of locals) said that they had chosen the track where they were contacted in order to avoid other people. Thirty-nine per cent of each group were carrying tents and 80% of locals were carrying cooking equipment although only 66% of visitors were doing so. When asked if carrying such equipment was to avoid using crowded huts or over-used facilities, over half said that it was, at least to some extent.

Given the emphasis accorded to crowding and reactions to it, it is not surprising that many people reacted to the presence of what they saw as too many others. In general terms, about a fifth of all respondents expected to see less people than they actually did; 34% would certainly have preferred to see fewer. This was particularly true of international visitors, who were most prevalent in the most popular locations; 25% would have preferred to have seen a few less and a further 16% a 'lot less' than they actually did.

As a consequence, 15% of those who felt that they had seen more people than they expected reported that they had become dissatisfied with their actual experience and about 16% said that they would choose somewhere else next time. When asked where they would go, everyone indicated that it would be to a more remote or a less well-known destination. Effectively, then, one in six has expressed the potential for both product shift and spatial displacement. Specifically, 16% of New Zealand residents said that they had changed their thoughts about the track, a figure that rose to 24% of the overseas component, representing considerable product shift and raising questions about the images that New Zealand projects of its back-country.

When asked if they took specific actions to avoid others, a third of both samples said that they did; the ways in which they did so are set out in Table

Table 5.6. Specific strategies to avoid too many people, per cent.

	Domestic	Overseas
Camped	40	32
Left early	10	18
Left late	8	16
Avoided crowded huts	13	7
Walked side tracks	4	14
Found secluded spots	7	9
Walked fast	5	7
Stayed by self	9	2
Walked in less popular directions	6	6
Allowed people to pass	5	6

5.6. Using a tent and camping was the most popular response, others left very early in the morning, so as to be first at the next hut and thus gain a bunk; others, principally the campers, stayed behind until all others had gone. These figures echo those found by Higham (1996) for international visitors only in the South Island.

Seasonal displacement is reflected in the fact that similar numbers from both samples, about 40%, said that there were tracks that they would not attempt to walk at the time that they were contacted. Three stood out in particular, namely the Milford, Routeburn and Abel Tasman tracks, the most popular of the Great Walks. The reasons given were anticipated crowding (69% of locals; 51% of visitors), the cost of hut fees (14 and 11%, respectively) or because they were too commercialized (4% in each case).

Satisfaction and mitigating strategies

In spite of the foregoing comments, most motivations to use the back-country were satisfied and overall levels of satisfaction remain high. Almost all of the sample report positive satisfaction or even extreme satisfaction, with New Zealanders marginally the more enthusiastic, again, perhaps, a reflection of their tendency to be away from the most crowded locations.

Previous studies (Kearsley and O'Neill, 1994) also offer the apparent paradox of considerable seeming dissatisfaction with particular attributes of an experience coupled with overall high levels of satisfaction with the total experience. This, in part, may reflect a subjugation of local irritations to contentment with a much more satisfactory whole. It might be that satisfaction and dissatisfaction are not necessarily polar opposites operating at just one level. Perhaps New Zealand's scenery is of such magnificence that some local discomfort is at present insignificant by comparison. This, clearly, is an area that requires further investigation.

Management of the Conservation Estate operates most specifically at the regional level (Corbett, 1995) with considerable variation among and within the Department of Conservation's 14 conservancies. In addition, the former New Zealand Forest Service had left a legacy of site-specific recreation plans and strategies. National strategies include the Great Walks concept, a move to increase revenue through hut fees and efforts to limit the impact of crowding on the most popular tracks (such as the Routeburn and Abel Tasman) through the rationing of hut passes. In addition, a national visitor strategy has now been produced (Department of Conservation, 1996). In the preparation of regional recreation strategies, the Recreation Opportunity Spectrum (ROS) approach (Driver and Brown, 1978; Clark and Stankey, 1979) has been used extensively and most of the Conservation Estate has been subject to ROS analysis. ROS is a spatial allocation process that enables managers to identify and provide for a diverse range of recreation opportunities through the manipulation of access, facilities and information. Its extension and modification through the carrying capacity concepts inherent in the Limits of Acceptable Change (LAC) process has not yet been implemented.

In recent years it has come to be recognized that much of the growth in forest recreation has been in the easily accessible 'front country' and this is not necessarily best served by the traditional provision of remote huts and tracks (Corbett, 1995). However, most of the visitors in this category nonetheless expect to encounter wilderness, at least to some extent. Balancing their needs with those of the 'purists' (Kliskey and Kearsley, 1993) who wish to encounter pristine wilderness has become a major task, as the crowding perceptions reported above illustrate. It is here that a new approach to management offers some prospects for accommodating various sets of wilderness values. This approach is based upon collecting and analysing personal perceptions of wilderness and using these to allocate potential users to forest environments that provide the wilderness context that they seek.

Personal images of wilderness

Just as attitudes to wilderness have varied over time by culture and society, so too have individual perceptions of what wilderness might be varied greatly. While wilderness environments have an objective reality as physical places, what makes that reality 'wilderness' rests very much upon personal cognition, emotion, values and experiences. As Stankey and Schreyer (1987) point out, a wilderness environment does not so much 'give' a wilderness experience as act as a catalyst for what are essentially inherent emotional states. Wilderness, then, has no commonly agreed physical reality, but it exists where personal cognitions say that it might be; different people perceive wilderness in different ways and in different places, but, for each of them, wilderness exists in that place, although others might vehemently disagree.

Many attempts have been made to explore the dimensions of the wilderness image (for example, Lucas, 1964; Hendee *et al.*, 1968; Stankey,

1971; Heberlein, 1973; Beaulieu, 1984). In New Zealand, Wilson (1979) showed that, while the general public and regular back-country users held similar views as to how wilderness might be described, seeing it as natural and unspoiled, wild, free and challenging, sacred, pure and exciting, the two groups had quite different views about what was permissible in a wilderness environment. Among trampers, the more purist did not believe it possible to have wilderness where there was any sign of people or their artefacts, whereas the public exhibited a much broader range of tolerance. Most of them, and, indeed, some trampers, believed that there was no inconsistency between a wilderness experience and the presence of such facilities as huts, tracks, swing bridges and even toilets and picnic sites. At the same time, there were clear limits as to what was acceptable, and vehicular access was strongly rejected, as was any evidence of overt commercialization. Thus, it appears that the highly purist required a pristine ecological wilderness, but that the majority could find wilderness values in places that had been part developed. This suggested that the saturation of pristine wilderness might be averted, as many found satisfaction in areas unacceptable to the purist minority.

The notion that wilderness could be encountered by various people in environments that were more or less developed was taken further (Kearsley, 1982; Shultis and Kearsley, 1988; Kearsley, 1990; Shultis, 1991; Higham, 1996; Kearsley, 1997). In various studies, wilderness users, the general public or international visitor users of the Conservation Estate were asked to state the extent to which they accepted various facilities (huts, tracks and bridges), characteristics (remoteness and solitude) or developments (exotic forests and mining) in wilderness areas. Kliskey and Kearsley (1993) show how responses to such a question may be used to group people into discrete purism classes and to plot the extent to which specific environments provide wilderness for various groups, using a GIS procedure. Kliskey's depiction of wilderness perceptions in what is now Kahurangi National Park is set out in Fig. 2. This work has been replicated for different groups of people and for different places (Kearsley *et al.*, 1997). It is clear that the demands for wilderness of the majority of forest users, and certainly the vast majority of the public, can be accommodated in environments that have been 'hardened' to minimize physical damage and provided with simple facilities that make recreation possible for substantial numbers of users.

Conclusions

New Zealanders' and overseas visitors' free and open access to native forests is greatly facilitated by the fact that they are almost entirely contained in almost all parts of the Conservation Estate. Nonetheless, their satisfaction has been affected by a large recent increase in overseas users, although that use is presently confined, for the most part, to the more popular and easier walking tracks. While this study has shown a high level of satisfaction with the experiences gained and the satisfaction of the motivations for a back-country experience, it is clear that there are significant perceptions of crowding, some environmental damage and noise pollution. It is equally clear that actual displacement, in various forms, has occurred and that there is a potential reservoir of more; there is a recognition by at least a significant minority that further visitor management controls are required as well. In the absence of large sample national studies in the past, it is unclear at what rate levels of crowding and associated phenomena are increasing, but it seems likely that they are doing so at least the rate of visitor increase. Further studies of this type may be necessary in the future so as to monitor patterns of change and their implications, and, indeed, a replication of that described here is scheduled for 2000–2002, while a similar survey of front-country users is close to completion.

New Zealand's native forests are now well protected against further logging or other commercial use but the war on introduced pests is far from over and, indeed, could well be lost, at least in some areas. The growing demands for forest recreation that have been encountered in recent years seem closely linked with a search for naturalness and wilderness values, which are incompatible with too high a level of use. It is suggested that the fact that wilderness means different things to different people means that simply designating a few pristine formal wilderness areas is not enough. A more humanistic definition, based upon human perceptions rather than the state of nature, will provide much wider opportunities for visitor satisfaction, through matching different wilderness perceptions with environments that most closely provide the experience and context desired.

New Zealand's remaining pristine native forests are preserved within the Conservation Estate, but they continue to face the consequences of two major invasions. One, the older, is of the animal species that have devastated both plant and bird life; the newer is the recent invasion of visitors from overseas who seek the wilderness and solitude that the forests can offer. As this chapter has attempted to show, traditional management methods are no longer adequate, and the consequences of largely tourist-induced crowding are beginning to impact significantly upon both traditional and recent users. The solution seems to lie in a greater understanding not of the forest ecology, but of the perceptions and expectations of the users themselves. The traditional wildlife focus of forest management must now be joined by a much stronger social scientific perspective, something that has not, as yet, sufficiently occurred.

References

Anderson, D. and Brown, P. (1984) The displacement process in recreation. *Journal of Leisure Research* 6, 61–73.

Beaulieu, J.T. (1984) Defining the components of the environmental image for use as a predictor of decision to participate. Unpublished PhD thesis, Utah State University, Logan, Utah.

Cameron, R.J. (1964) Destruction of the indigenous forests for Maori agriculture during the nineteenth century. *New Zealand Journal of Forestry* 9, 98–109.

Clark, R.N. and Stankey, G.H. (1979) The Recreation Opportunity Spectrum: a framework for planning, management and research. USDA Forest Research Paper PNW-98.

Corbett, R.A. (1995) Managing Outdoor Recreation. In: Devlin P.J., Corbett, R.A. and Peebles, C.J. (eds) *Outdoor Recreation in New Zealand.* Department of Conservation, Wellington.

Cumberland, K.B. (1965) *The Moahunter.* Whitcombe and Tombs, Christchurch.

Department of Conservation (1995) *Visitor Strategy.* Department of Conservation, Wellington.

Driver, B.L. and Brown, P.J. (1978) The opportunity spectrum in outdoor recreation supply inventories: a rationale. *Proceedings of the Integrated Renewable Resource Inventories Workshop.* USDA Forest Service General Technical Report RM-55.

Fitzharris, B.B. and Kearsley, G.W. (1987) Appreciating our high country. In: Holland, P.G. and Johnston, W.B. (eds) *Southern Approaches.* New Zealand Geographical Society, Dunedin, New Zealand, pp. 197–218.

Glacken, C.J. (1967) *Traces of the Rhodian Shore: Nature and Culture in Western Thought from Ancient Times to the End of the Eighteenth Century.* University of California Press, Berkeley.

Hall, C.M. (1992) *Wasteland to World Heritage: Preserving Australia's Wilderness.* Melbourne University Press, Carlton.

Heberlein, T.A. (1973) Social psychological assumptions of user attitude surveys: the case of the Wildernism Scale. *Journal of Leisure Research* 5, 18–33.

Hendee, J.C., Catton, W.R., Marlow, L.D. and Brockman, C.F. (1968) Wilderness Users in the Pacific Northwest: Their characteristics, values and management preferences. Research Paper PNW-61, US Department of Agriculture Forest Service, Pacific Northwest Forest and Range Experiment Station, Portland, Oregon.

Higham, J.E.S. (1996) Wilderness Perceptions of International Visitors to New Zealand. The perceptual approach to the management of international tourists visiting wilderness areas within New Zealand's Conservation Estate. PhD thesis, University of Otago, Dunedin.

Higham, J.E.S. and Kearsley, G.W. (1994) Wilderness Perception and its implications for the management of the impacts of international tourism on natural areas in New Zealand. In: *Tourism Down-under: A Tourism Research Conference, 6–9 December, 1994.* Palmerston North: Department of Management Systems, Massey University, pp. 505–529.

Kearsley, G.W. (1982) *Visitor Survey of Fiordland National Park.* Lands and Survey Department, Wellington, New Zealand.

Kearsley, G.W. (1990) Tourist development and wilderness management in Southern New Zealand. *Australian Geographer* 21, 127–140.

Kearsley, G.W. (1995) Recreation, tourism and resource development conflicts in Southern New Zealand. *Australian Leisure* 26–30.

Kearsley, G.W. (1996) The impacts of tourism on New Zealand's back country culture. In: Robinson, M., Evans, N. and Callaghan, P. (eds) *Tourism and Cultural Change*. Centre for Travel and Tourism, University of Northumbria, Newcastle, pp. 135–146.

Kearsley, G.W. (1997) Managing the consequences of over-use by tourists of New Zealand's conservation estate. In: Hall, C.M., Jenkins, J. and Kearsley, G.W. (eds) *Tourism, Planning and Policy in Australia and New Zealand; Issues, Cases and Practice*. Irwin, Sydney, pp. 87–98.

Kearsley, G.W. and Higham, J.E.S. (1996) Wilderness and back country motivations and satisfaction in New Zealand's natural areas and conservation estate. *Australian Journal of Leisure and Recreation* 8, 30–34.

Kearsley, G.W. and Higham, J.E.S. (1997) *Management of the Environmental Effects Associated with the Tourism Sector: Review of Literature on Environmental Effects*. Parliamentary Commissioner for the Environment, Wellington.

Kearsley, G.W. and O'Neill, D. (1994) Crowding, satisfaction and displacement: the consequences of the growing tourist use of Southern New Zealand's conservation estate. In: Ryan, C. (ed.) *Tourism Down Under*. Massey University, Palmerston North, New Zealand, pp. 171–184.

Kearsley, G.W., Kliskey, A.D., Higham, J.E.S. and Higham, E.C. (1997) Different people, different times: different wildernesses. In: Higham, J.E.S. and Kearsley, G.W. (eds) *Trails in the Third Millenium*. Centre for Tourism, University of Otago, Dunedin, pp. 197–214.

Kearsley, G.W., Coughlan, D.P., Higham, J.E.S., Higham, E.C. and Thyne, M.A. (1998) Impacts of tourist use on the New Zealand backcountry. Research paper No. 1, Centre for Tourism, University of Otago, Dunedin, New Zealand.

Keogh, C. (1991) Routeburn Track Market Study. Unpublished MBA dissertation, University of Otago, Dunedin, New Zealand.

Kliskey, A. and Kearsley, G.W. (1993) Mapping multiple perceptions of wilderness in North West Nelson, New Zealand: a geographic information systems approach *Applied Geography* 13, 203–223.

Lucas, R.C. (1964) Wilderness perception and use: the example of the Boundary Waters canoe Area. *Natural Resources Journal* 3, 394–411.

McGlone, M.S. (1983) Polynesian deforestation of New Zealand – a preliminary synthesis. *Archaeology in Oceania* 18, 11–25.

Memon, P.A. and Wilson, G.A. (1993) Indigenous forests. In: Memon, P.A. and Perkins, H.C. (eds) *Environmental Planning in New Zealand*. Dunmore Press, Palmerston North, pp. 97–119.

Molloy, L.F. (1983) Wilderness recreation – The New Zealand Experience. In: Molloy, L.F. (ed) *Wilderness Recreation. Proceedings of the FMC 50th Jubilee Conference on Wilderness*. Rotoiti Lodge, Nelson Lakes National Park, August 1981.

Moore, K. (1995) Understanding the individual recreationist: from motivation to satisfaction. In: Devlin, P.J., Corbett, R.A. and Peebles, C.J. (eds) *Outdoor Recreation in New Zealand*. Department of Conservation and Lincoln University, Wellington.

Nash, R. (1982) *Wilderness and the American Mind*, 3rd edn. Yale University Press, New Haven.

New Zealand Tourism Board (1996) *International Visitor Survey 1995/6*. NZTB, Wellington.

Nielson, J.M. and Endo, R. (1977) Where have all the purists gone? An empirical examination of the displacement hypothesis. *Western Sociological Review* 8, 61–75.

Oelschlager, M. (1991) *The Idea of Wilderness: From Prehistory to the Age of Ecology*. Yale University Press, New Haven and London.

Shultis, J.D. (1991) Natural environments, wilderness and protected areas: an analysis of historical western attitudes and utilisation, and their expression in contemporary New Zealand. Unpublished PhD thesis, University of Otago, Dunedin, New Zealand.

Shultis, J.D. and Kearsley, G.W. (1988) Environmental perception in protected areas. In: *Proceedings of the Symposium on Environmental Monitoring in New Zealand*. Department of Conservation, Wellington, pp. 166–177.

Stankey, G.H. (1971) The perception of wilderness recreation carrying capacity: a geographic study in natural resource management. Unpublished PhD thesis, Michigan State University, East Lansing, Michigan.

Stankey, G.H. and Schreyer, R. (1987) Attitudes towards wilderness and factors affecting visitor behaviour: A state of knowledge review. *Proceedings, National Wilderness Research Conference: Issues, State of Knowledge and Future Directions*. General technical Report INT-220. Intermontane Research station, Ogden, Utah, pp. 246–293.

Ward, J.C. and Beanland, R.A. (1996) *Biophysical Impacts of Tourism*. Centre for Resource Management. Information paper No. 56, Lincoln University, Lincoln, New Zealand.

Wilson, G.A. (1991) The urge to clear the bush – a study on the nature, pace and causes of native forest clearance on farms in the Catlins district (South East South Island New Zealand). Unpublished PhD thesis, University of Otago, Dunedin, New Zealand.

Wilson, M.L. (1979) Dimensions of the wilderness image. Unpublished BA (Hons) dissertation, Department of Geography, University of Otago, Dunedin, New Zealand.

6

A Review of Ecology and Camping Requirements in the Ancient Woodlands of the New Forest, England

David Johnson and Angela Clark

Introduction

The New Forest, in central southern England, is of international nature conservation importance. It is the most ecologically important assemblage of lowland heath and ancient semi-natural pasture woodland in northern Europe. This is reflected in a range of established and proposed nature conservation designations. In 1987, 27,734 ha of the New Forest were re-notified as Sites of Special Scientific Interest (SSSIs)[1] by English Nature. Parts of the New Forest have been recommended as a Ramsar site, Special Protection Area (SPA) and Special Area of Conservation (SAC). It is also proposed as a World Heritage Site. A £5 million European Union *LIFE*-Nature grant, awarded in 1997, is being used to remove alien trees, re-introduce pollarding, control bracken, clear rhododendron, repair erosion and effect habitat restoration (Forestry Commission, 1997a).

The New Forest contains significant areas of semi-natural ancient woodland (Tubbs, 1968; Peterken, 1993; Peterken *et al.*, 1996). These unenclosed woods have been subjected to grazing and browsing by deer and domestic stock depastured on the Forest by commoners since Anglo Saxon times (Rackham, 1980). This type of management has resulted in their development into pasture-woodland. There are three basic types of pasture-woodland, namely Wooded Commons, Forests and Parks. The Ancient and Ornamental[2] woods of the New Forest are of the Forest type (Colebourn, 1983).

The New Forest Heritage Area has National Park status for planning purposes, and is under substantial recreational and development pressures. It is a major UK tourist destination and an important venue for both formal and

(eds Xavier Font and John Tribe)

informal countryside recreation. In 1996–1997 the Forestry Commission estimated visitor numbers to be in excess of 7 million. Camping has occurred in the New Forest on an informal, *ad hoc* basis since the turn of the century. However, by the mid 1960s, pressure of numbers coupled with lack of facilities, led to recommendations that the practice of wild-camping, where campers had unrestricted access to all crown lands of the New Forest, should be drastically restricted; vehicular access to the forest constrained by ditches and barriers; and campsites and car parks with appropriate facilities established. As a result the first formal campsite was established on the site of a World War II airfield at Holmsley in 1964. A policy of dispersal was advocated to distribute recreational pressures evenly throughout the New Forest car parks and campsites.

Some attention was paid to the questions of site suitability when locating car parks and campsites, in terms of a particular habitat's ability to withstand the pressures of recreation. However, this tended to be overshadowed by the Forestry Commission's policy to establish campsites in locations already attracting a large volume of campers, generally close to villages and main roads, and frequently in the picturesque Ancient and Ornamental woodlands. Development of these campsites was subject to review in 1976. Evidence from aerial photographs and ground surveys highlighted significant adverse environmental impacts at most sites. The review recommended modification of the design and management of some campsites by construction of gravel access roads within the campsites, provision of more facilities, and stricter controls on activities such as lighting campfires (New Forest Technical Review Group, 1976). The Forestry Commission's Management Plan for the period 1982–1991 put some of these recommendations into practice, determining that there would be no new facilities but that existing facilities would be maintained and renewed. A second review in 1988 recommended a reduction in overall pitch numbers within the New Forest to a total of 3200 (excluding overflow areas used at times of peak demand). This reflected the drop in demand recorded from the peak of 853,900 camper nights in 1978 to approximately 750,000 camper nights over the subsequent years up to 1987.

The economic importance of campsites to the Forestry Commission is demonstrated by the high proportion they contribute to the Commission's annual New Forest income. In the financial year 1996/97 camping revenue totalled £1,582,000; almost half the total revenue for the New Forest of £3,039,000. In the same year the Commission spent £1,168,000 on camping-related management. The campsites are also of significant economic importance to the adjacent villages. One local business association estimates that 75% of their annual revenue is derived from the campers holidaying at the three nearby Forestry Commission campsites.

Importance of Ancient Wood-pasture

Ancient semi-natural traditional coppice and wood-pasture stands dating from the Middle Ages support a much greater diversity of species than more recent and plantation woodland (Peterken, 1993). This can be explained as follows:

1. long establishment allows for the development of microhabitats which favour specialist niche species and the opportunity for the chance arrival of many species;
2. ancient woods comprise mainly indigenous tree species which have more species of vertebrates, invertebrates and epiphytic lichens, bryophytes and fungi associated with them than non-indigenous species;
3. these woodlands also have a high humidity and tend to contain a significant amount of dead and decaying wood, deep soils and leaf litter which provides habitat for a rich diversity of birds, invertebrates and cryptogams (Rose and James, 1974; Ratcliffe, 1977; Peterken, 1993);
4. due to the absence of dense understorey, particularly in ancient wood-pasture, light levels are high, encouraging some species of lichen. Certain epiphytes have been used as indicator species to construct Indices of Ecological Continuity. Open spaces are also a characteristic of ancient woodlands. Mosaics of light and shade add to a rich assortment of spaces and holes for nesting, shelter, hiding and sunning;
5. larger, relatively undisturbed ancient woods provide breeding habitat for sensitive creatures such as deer and woodcock.

Consequently ancient woodlands are highly valued by nature conservationists (Spencer and Kirby, 1992; English Nature, 1998). Wood-pasture, as found in the New Forest, is rarer than ancient coppice and is notable for containing more mature trees and dead wood than ancient coppice woodland. Harding and Rose (1986) suggested that protecting existing lowland wood-pasture sites should be a top priority for woodland conservation.

Even if it is accorded this priority, increasing isolation of ancient woodland stands threatens rare ground flora species and invertebrates with poor mobility. The Ancient Woodland Inventory Project, undertaken in the 1980s, revealed that whilst ancient woodland was widespread it only accounted for 2.6% of the land surface of England. Furthermore, 83% of sites were less than 20 ha and less than 2% more than 100 ha. The New Forest is one of the noticeable concentrations (Saunders, 1993).

Balancing Recreational and Ecological Interests

Camping impacts

Lane and Tait (1990) summarized the impacts of recreation on woodland. At formal campsites, cars, caravans, bicycles and campers' feet cause soil compaction, erosion, alterations to natural drainage and damage to trees.

Trampling, for example, results in maceration and removal of leaf litter and a reduction in the depth of organic soil layers. It can also damage or destroy ground flora, further reducing soil porosity. Absence of ground flora exposes the area to erosion, especially since rainfall cannot easily penetrate the compacted soil and the increased runoff will carry away soil loosened by feet and vehicles (Liddle, 1997).

Emissions from cars, camping stoves and barbecues may also cause ecological damage. Little has been published in this area but there are indications that lichens in particular are extremely sensitive to the emissions from kerosene-burning appliances (Rose, personal communication,1998).

When access to a site is improved by the installation of hard or raised roads and tracks, the passage of vehicles over them during dry conditions causes dust which resettles on adjacent ground flora and tree trunk surfaces. This is believed to cause eutrophication and changes the pH of the soil and tree trunk surfaces. These tracks, roads and other surfacing can also affect the drainage of the site causing localized water-logging and/or water starvation. Dog urine can also increase eutrophication.

The timescale which needs to be considered when considering impacts on ancient woodland is a complicating factor. By the time very mature trees start to show signs of pressure it may be too late to save them.

Recreation planning

Recreation planning strategies, based on ascertaining the optimum visitor numbers, which in turn should influence campsite design, pitch numbers and length of season, are often based on the determination of recreational carrying capacity (Glyptis, 1991; Glasson *et al.*, 1995). However, there are conceptual problems with establishing carrying capacity for camping (i.e. variety of type, intensity and seasonality of camping) and, for ancient pasture-wood, ecological carrying capacity in particular is low due to the sensitivity of many of its associated flora.

A strategic decision either to concentrate or disperse camping activity can be made. For some recreational activities, such as water-based recreation, spatial and temporal zoning plans have also proved a useful means of restricting visitor pressure. Within a campsite Beazley (1969) advocated simplistic zoning, allocating different zones for tents and caravans and restricting cars to specific areas, thus reducing the requirement for hardened surfaces. For this to work, campsite design must take into account camper behaviour.

Sidaway (1993) champions establishing *Limits of Acceptable Change*, an approach developed by the US Forest Service, whereby use levels are agreed by all parties. The advantage of this method is its collaborative approach, involving all stakeholders advised by experts.

Self-regulation is also important, used in conjunction with one or a combination of any of the above. In this respect informing and empowering campers is necessary. To that end, in the UK, the Camping and Caravan Club has its own Environmental Code and a Good Practice Guide has been produced for Holiday Caravan Parks.

Practical management

Practical visitor management can both reduce/remediate or exacerbate the ecological impacts of camping. For New Forest campsites this work includes:

1. annual tree safety surveys (both within campsites and 20 m outside the perimeter);
2. rotavation of compacted soil to improve surface drainage;
3. localized additions of topsoil and reseeding with special indigenous grass seed mixture;
4. placement of physical barriers, such as dragons' teeth and post and rail barriers, to restrict visitors to prescribed sites.

Hollands Wood: Case Study

Background

Hollands Wood is one of the New Forest's Ancient and Ornamental pasture-woods. The Revised Index of Ecological Continuity gave this site a score of 100 out of a possible 150, which demonstrated its stature and importance in conservation terms (Peterken, 1993).

In 1970/71 the Forestry Commission established a 750-pitch campsite at Hollands Wood at a cost of £40,000. In line with the 1988 New Forest Review, pitches were reduced to 600 in 1990. A further reduction in pitch numbers to 570, together with access changes and facility upgrades, was proposed by the Forestry Commission in the latest forest-wide review of camping (Forestry Commission, 1995). Nevertheless, Hollands Wood is the most popular and the most expensive of the nine Forestry Commission campsites in the New Forest. Its revenue alone regularly accounts for almost one-third of the total revenue generated by all nine of the New Forest campsites.

Over the 28 years of its operation to date, the campsite has had a significant negative environmental impact as a result of disturbance and alterations to the woodland soil structure and composition. In 1975/76 some of the older trees at Hollands Wood were felled, both for safety reasons and to increase space on the campsite.

A report by the New Forest Association and Hampshire Wildlife Trust (Cox and Rose, 1996) stated that since its establishment as a campsite:

1. 84% of the mature trees have been lost, reducing canopy cover by 50%;
2. 76% of the site can be classified as heavily disturbed ground;
3. 16% of the site has been covered by roads, tracks and buildings;
4. there has been a significant reduction in the variety and distribution of lichen flora; and
5. the site has suffered a substantial loss of landscape quality.

The report concluded that Hollands Wood could not be sustained as an area of Ancient and Ornamental woodland with the current level of camping intensity and recommended that the campsite be relocated in order to allow the woodland to recover for a period of 50–100 years.

Partially in response to the New Forest Association report, the Forestry Commission produced 'A Framework for Recreation', which set out the Commission's commitment to review the size and location of campsites, consider closure or relocation of sites in sensitive locations, carry out environmental impact assessments prior to undertaking potentially damaging operations and to repair and prevent recreation-related erosion (Forestry Commission, 1997b).

Environmental appraisal 1998

During the autumn/winter of 1998 a detailed site survey of the Hollands Wood site was undertaken by Southampton Institute. The campsite was compared with similar adjacent areas of Ancient and Ornamental woodland within which camping is not permitted and which were sufficiently divorced to avoid secondary disturbance. The Geographic Information System (GIS) package MapInfo Version 5 was used to present the results. Within the campsite the survey highlighted:

1. significant impoverishment of ground flora caused by both compaction, due to vehicles and trampling, and shading from tents and caravans (0.36 ha out of the total campsite area of 1.44 ha or 25% of the site was classified as bare ground);
2. increased tree surgery of mature trees for safety purposes, reducing their ecological value;
3. little natural regeneration;
4. greater openness, probably as a result of felling operations in 1975/76;
5. reductions in the amount of deadwood and woodland debris which can be attributed to pre-camping season maintenance and removal of wood by visitors; and

6. impacts of campsite infrastructure (gravel tracks, ablution blocks, waste reception facilities).

To ameliorate these problems the following design and management changes were proposed to the Forestry Commission:

1. removal of all gravel tracks and their replacement with chipped bark surfacing;
2. restrictions on car and caravan access;
3. introduction of a pricing structure which favours tenting rather than caravans;
4. provision of trolleys, similar to those currently provided at marinas, to transport campers equipment from vehicle areas to tent pitches;
5. introducing campsite restrictions on the use of kerosene gas-powered equipment, encouraging the use of butane equipment in its place;
6. introducing a ban on dogs in line with existing bans at two other Forestry Commission campsites at Ashurst and Denny Wood;
7. protecting areas around the bases of more ecologically important trees using logs as natural barriers;
8. providing habitats for invertebrates and food sources for birds by relaxing forest hygiene within the campsite, leaving deadwood and other forest debris *in situ*;
9. a tree/shrub replacement/replanting programme; and
10. providing more interpretation and educational materials relating to the ecological importance of Hollands Wood and the rarity and significance of ancient pasture-woodland, in order to raise awareness of the fragility of the wood and to encourage responsible and environmentally friendly visitor behaviour.

These recommendations are in line with the new government forest strategy (HM Government, 1999) which targets the Biodiversity Action Plan – listed habitats, pledges to introduce long-term management plans for ancient semi-natural woods and promotes the environmental benefits of trees and woodlands. The study also prompted a more detailed review of recreational pressures on vulnerable habitats in the New Forest using a similar methodology to that employed for Hollands Wood. In this study (Clark, 1999), 12 sites subject to heavy recreational use were individually mapped and photographed. Information on site conditions, habitat types, recreation types and the existence of more robust potential alternative sites and recreational features were entered into GIS MapInfo. A recreation pressure map was also produced by entering information for all campsites, cycle hire shops, riding schools, livery stables, Forestry Commission car parks and bridges, roads and population densities (by parish) in and around the New Forest. Of the 12 sites selected, 10 showed significant signs of ecological damage related to various recreational uses. On a percentage basis 9.7% of heath/mire complexes; 5.6% of grassy 'greens', bracken and gorse stands found around the heath/mire complexes; and 0.7% of woodland were cause for concern.

Heath/mire complexes were shown to be particularly at risk from over-use by horse riding, cycling (despite existing prohibitions) and walking.

Conclusions

Evidence from the New Forest suggests that controlling access and prohibiting wild-camping, a policy initiated in the 1960s, has been instrumental in preserving the broad fabric of the New Forest at the expense of a number of 'honeypot' sites. Potential impacts of camping include erosion, soil compaction, tree damage, wildlife disturbance, trampling, accidental fires, littering and vandalism.

As the timescale required for the re-establishment and/or regeneration of ancient woodland is measured in centuries, strategies for the conservation of existing examples of this habitat are essential. All woodland has a high perceptual carrying capacity. However, in the case of the New Forest, recreational planning decisions of the early 1970s ignored the low ecological carrying capacity of Ancient and Ornamental woodland. There is now a general consensus that in ancient woods commercial considerations should be subordinated to those of nature conservation (Fuller and Peterken, 1995). Conversely, recent woodlands, particularly monocultures of spruce and pine, are of more limited nature conservation interest. Indeed, leisure developments such as Center Parcs at Elveden, Longleat and Sherwood Forests in the UK, located within coniferous plantations, have demonstrated significant biodiversity gains without compromising visitor enjoyment. Clearly if concentration strategies for woodland recreation are adopted it is important to ensure that the honeypot site is either robust enough to cope with the recreation pressures or of little or no conservation value.

Much of the literature on the management of camping in woodland assumes that the woodland can be manipulated in the interests of both conservation and recreation by, for example, creating glades and rides. If Ancient and Ornamental pasture-woods are subjected to this type of management regime they lose a substantial amount of their conservation value and their future integrity may well be compromised.

This study of Hollands Wood, a 750-pitch Forestry Commission campsite in the heart of the New Forest, illustrates the scale of potential adverse impacts on semi-natural ancient woodland. Hollands Wood is important to the local economy but internationally important ecologically, particularly for its lichen flora and because it is believed to support a colony of rare Bechstein's bats (*Myotis bechstein*). Camping has seriously damaged the site, necessitating the loss of many of the older trees, causing soil compaction and erosion, and reducing landscape quality. If campsites have been established in ecologically sensitive and important woodlands, as is the case at Hollands Wood, it is important to give detailed attention to the design and management of the campsite. Visitor facilities and behaviour should be manipulated to the benefit of the woodland, rather than manipulating the woodland to the benefit of the visitors. As shown by the Hollands Wood case study, a number of simple management changes can reduce ecological impacts. These include the redesign of facilities, access restrictions, track resurfacing, bans on dogs and kerosene-burning equipment, interpretation and education. In this respect the case study provides lessons for the planning and management of other campsites both within and beyond the New Forest.

Agencies involved in promoting the New Forest as a visitor destination are aware of their obligations under Local Agenda 21. A strategic partnership approach to forest tourism has been established (New Forest District Council, 1998). However, if it is accepted that the survival and protection of rare habitat such as Ancient and Ornamental woodland is the over-riding priority, then the Forestry Commission should invest in a consensus-building approach (Sidaway, 1998) targeted at relocating campsites established within ancient woodland. With a campsite such as Hollands Wood, which is hugely popular and a significant local economic generator, this process will inevitably be contentious.

Acknowledgements

The authors would like to thank Ted Johnson, Chairman of the New Forest Committee, and Dr Roger Sidaway, Research and Policy Consultant, for their comments on an earlier draft of this paper.

Notes

[1] SSSIs were established in the UK as a nature conservation designation under the National Parks and Access to the Countryside Act 1949, and subsequently re-notified (resurveyed and reassessed in terms of scientific importance) or initiated under the Wildlife and Countryside Act 1981 and Amendment 1985.

[2] Ancient and Ornamental was first used as a term to describe these historic pasture woodlands in Section 8 of the New Forest Act 1877.

References

Beazley, E. (1969) *Designed for Recreation.* Faber & Faber, London.

Clark, A. (1999) Environmental assessment of vulnerable habitats: a review of recreation pressures in the New Forest. Report to the Forestry Commission Ref: LIFE/98/L30A2T/01. Unpublished.

Colebourn, P. (1983) *Hampshire's Countryside Heritage 2: Ancient Woodland.* Hampshire County Planning Department, Winchester.

Cox, J. and Rose, F. (1996) *A Preliminary Assessment of Proposed Changes in Camping and Car Parking Provision in the New Forest.* Jonathon Cox Associates, Winchester.

English Nature (1998) *Management Choices for Ancient Woodland: Getting it Right.* English Nature, Peterborough.

Forestry Commission (1995) *New Forest Camping: A Review and Options for the Future.* Forestry Commission, Lyndhurst.

Forestry Commission (1997a) *Life in the New Forest 1997.* Forestry Commission, Lyndhurst.

Forestry Commission (1997b) *A Framework for Recreation in the Crown Lands of the New Forest.* Forestry Commission, Lyndhurst.

Fuller, R.J. and Peterken, G.F. (1995) Woodland and scrub. In: Sutherland, W.J. and Hill, D.A. (eds) *Managing Habitats for Conservation.* Cambridge University Press, Cambridge, pp. 327–361.

Glasson, J., Godfrey, K. and Goodey, B. (1995) *Towards Visitor Impact Management.* Avebury, Aldershot.

Glyptis, S. (1991) *Countryside Recreation.* Longman, Essex.

Harding, P.T. and Rose, F. (1986) *Pasture-Woodlands in Lowland Britain.* Institute of Terrestrial Ecology, Huntingdon.

HM Government (1999) *Forestry Strategy – A New Focus for England's Woodlands.* HMSO, London.

Lane, A. and Tait, J. (1990) *Practical Conservation: Woodlands.* Hodder & Stoughton, London.

Liddle, M. (1997) *Recreation Ecology.* Chapman & Hall, London.

New Forest District Council (1998) *Our Future Together: A Tourism and Visitor Management Strategy for New Forest District.* New Forest District Council, Lyndhurst.

New Forest Technical Review Group (1976) Progress report on the Implementation of Conservation Measures 1972 – 76. Forestry Commission, Lyndhurst.

Peterken, G.F. (1993) *Woodland Conservation and Management,* 2nd edn. Chapman & Hall, London.

Peterken, G.F., Spencer, J.W. and Field, A.B. (1996) *Maintaining the Ancient and Ornamental Woodlands of the New Forest.* Forestry Commission, Bristol.

Rackham, O. (1980) *Ancient Woodland: its History, Vegetation and Uses in England.* Edward Arnold, London.

Ratcliffe, D.A. (1977) *A Nature Conservation Review.* Cambridge University Press, Cambridge.

Rose, F. and James, P.W. (1974) Regional Studies of the British Lichen Flora 1: The corticlous and lignicolous species of the New Forest. *Lichenologist* 6.

Saunders, G. (1993) Woodland conservation in Britain. In: Goldsmith, F.B. and Warren, A. (eds) *Conservation in Progress.* Wiley, Chichester, pp. 67–95.

Sidaway, R. (1993) Sport, recreation and nature conservation: developing good conservation practice. In: Glyptis, S. (ed.) *Leisure and the Environment – Essays in honour of Professor J.A. Patmore.* Belhaven Press, London, pp. 163–173.

Sidaway, R. (1998) *Good Practice in Rural Development No. 5: Consensus Building.* Published by The Scottish Office Agriculture Environment and Fisheries Department for Scottish National Rural Partnership.

Spencer, J.W. and Kirby, K.J. (1992) An inventory of ancient woodland for England and Wales. *Biological Conservation* 62, 77–93.

Tubbs, C.R. (1968) *The New Forest: An Ecological History.* David & Charles, Newton Abbot.

7

Ecotourism on the Edge: the Case of Corcovado National Park, Costa Rica

Claudio Minca and Marco Linda

Introduction

Tourism can represent an extraordinary and sometimes unexpected source of development for the local communities in forested areas. Tourism can create new jobs, stimulate migration flows, and introduce new social dynamics within a local system. At the same time, if properly developed, tourism is capable of supporting a new awareness and new representations of the local culture that can be revitalized by the interest of the tourist. Tourism also introduces new forms of territorial organization with their accordant hierarchies, core-periphery logics and modalities of spatial segregation. Local development is, therefore, strongly influenced by the impacts of tourism and tourist spatialities.

This chapter is an attempt to develop a theoretical approach to describe the role of tourism for local communities and their territorialities, with a particular focus on forested areas. In particular, we shall query the explanatory potential – as well as the limits – of such a theoretical framework by examining the case of Costa Rica's Corcovado National Park. We thus begin with a brief introduction detailing Costa Rica's long tradition of environmental protection, then progressing to some considerations on the role of tourism in this Central American country. These two rather general parentheses will help us 'contextualize' our examination of the Osa Peninsula and Corcovado National Park. Our case study rests upon an analysis of the pertinent social agents and their associated territorial processes operating within the environment of the Peninsula and the National Park. The above noted analytical framework is informed, in large part, by approaches formulated within a geographical interpretation of systems theory through the analysis of the concepts of region and regionalization (Vallega, 1982; Turco, 1984; Vallega, 1995); approaches which have, as their scope, the goal of identifying the relationships between tourist development and the multitude of other processes which have contributed to the forging of the territorialities of the Peninsula and the Park, attempting to identify both their principal areas of conflict – as well as points of potential synergy.

Tourist Territoriality and Local Development: a Theoretical Approach

The theoretical approach we will develop aims to identify and describe the most significant processes influencing the territorial evolution of a local system. As noted previously, our model of tourist territoriality is grounded within a particular, 'geographical' elaboration of systems theory which seeks to codify the relationship between tourism and regional/local development: in the paragraphs that follow, we shall delineate this perspective along with its application in a very specific geographical context.

Certainly, our approach constitutes but one possible way to theorize the many intriguing questions raised by tourism's various impacts. As stressed above, our model emerges out of the systemic discussion of the concepts of region and regionalization developed by a group of Italian geographers during the past decade (see, above all, Vallega, 1982; Turco, 1984; Vallega, 1995). In order to avoid any terminological confusion it is, perhaps, appropriate to begin our theoretical elaboration by defining what we shall intend by a 'territorial system'. Following the theorization of the above-mentioned Italian school, a territorial system consists of a group of related territorial elements whose relationships are particularly strong and functionally significant. This web of relationships gives life to distinct organizational processes: it is thanks to these latter that it is possible to make a distinction between what is part of the system and what lies outside of it. The ultimate scope of these processes is to achieve and maintain a certain degree of autonomy for the territorial system. To achieve such autonomy, however, the territorial system must be 'properly structured': thus, regulated by a legitimized normative code. The legitimization of certain consolidated rules, which are thought to preserve the organization and the autonomy of the system is, therefore, one of the consequences of the social dialectic operative within the system. In fact, the interpretation and the definition of the normative code that (spatially) structures the system are the result of the dialectic between different social agencies and individuals who recognize that particular piece of land as part of 'their' territorial system. It must be clear, however, that not every area and every community is part of some territorial system. A territorial system is, therefore, an area characterized by a distinct, 'special' identity that derives from its normative code and from the resultant spatial structuration.

What, then, is the role of tourism in the development/structuring of territorial systems? Certainly, we can assert that the tourist is particularly attracted by the most evident and spectacular forms of the 'special' identity that territorial systems exhibit, largely because one of the motivations of her/his departure is the search for a break of continuity with everyday life. This rupture relies precisely on the possibility of visiting different cultural contexts, diverse natural and human landscapes, experiencing liminal environments and ephemeral social relationships. For this reason, the specificity of a well-structured territorial system and the geographical signs of its identity are powerful factors of attraction for the tourist. The relationship between the territorial processes associated with tourist development and the territorial system is thus quite complex.

To better formulate this relationship, it is necessary to further elaborate the theoretical framework described above. As we noted before, it is the organization of the system which frames the distinction between the system itself and the external environment – an assumption that emphasizes the relevance of the particular mechanisms that govern the system's external relationships. According to Turco (1988), territorial systems are governed by both 'auto-centric' territorial processes (the expressions of internal social and economic dialectics), as well as by 'hetero-centric' territorial processes (the result of 'exogenous' projects *for* that territory). Following Turco's theorization, tourist developments fall within the latter category. Tourism, in fact, is seen as the cultural product of the tourists' society of departure that can bring about a potential functional re-definition and territorial reorganization of certain local systems. In other words, tourist territoriality is the spatial expression of the way 'we' see 'them' and, inescapably, of the way 'they' see 'us'. The impacts of tourism on the autonomy and stability of local systems are thus worthy of attention. We should not forget, however, that every local territorial system subject to a tourist re-territorialization is, obviously, also characterized by pre-existent territorialities as well as other coexisting territorial processes.

Briefly, thus, the existence of a territorial system is based upon the stability of its normative code and social structure. Such stability requires the consolidation of certain spatial practices and an adequate spatial structure, presumably framed so as to support and legitimize the social structure. The system's very existence, however, is also dependent upon its metabolization of external impulses. Thus, the adaptation to any kind of innovation (new information, border pressures, the introduction of new territorial processes, etc.) is capable of provoking either a reinforcement of the system, if properly absorbed, or its disintegration/weakening, if the innovation disrupts the normative code that rules the system and supports its structure.

Tourist territorialization can have quite devastating results for the system when it is radically imposed by external forces; this is certainly the case when state governments accept and promote the spatial reorganization of peripheral areas following plans laid down by international tourist corporations. The construction of five-star mega-resorts and large hotels designed to host international clients very often implies a marginalization of pre-existing territorial activities. Very rarely is the local community involved in such a process of territorial reorganization; as a consequence, the local system's social and spatial structure tends to be overwhelmed by the new hetero-centric territorial order.

Taking the above into account, the geographical significance of tourist development might successfully be read within a systemic framework. If tourism is integrated into a well-structured economic and social web, it is likely to become a powerful and attractive complementary strategy of development for the system. On the contrary, when it becomes the dominant if not the only territorial process driving the local system, it can give place to dangerous dependencies upon external markets and external projects; if this is the case, the system tends to lose its autonomy, thus jeopardizing its existence as a system. In other words, the system (to survive as a system) has to be *cognitively open* (as innovation can also entail opportunities), though at the same time *normatively closed*, which signifies that its 'base' rules and structures should be immune from the deleterious influence of external factors.

Within geographical literature, it is the 'region' that has long been conceptualized as representing the territorial system par excellence: a distinct system characterized by a very strong autonomy and dominated by auto-centric processes (Turco, 1984). Assuming the region as just such a territorial system,

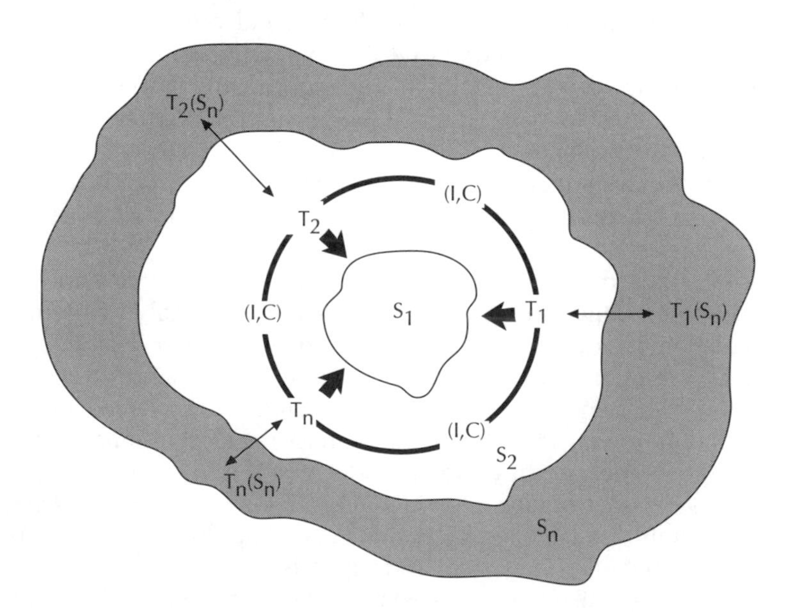

Fig. 7.1. A territorialization model. $T_{1,2,\ldots,n}$ = Territorialization processes; $S_{1,2,\ldots,n}$ = territorial system; I = interaction; C = conflict. Source: Minca and Draper (1997).

we can note that tourism can represent a potential factor of regionalization (that is, of the reinforcement of the regional entity) only if integrated in a synergistic relationship with the regional web; it is, however, a dramatic factor of de-regionalization when it jeopardizes the regional identity and autonomy by imposing exogenous rules and projects.

Our aim in this paper, then, is to delineate a possible methodology for the examination of: (i) the sustainability of tourist territorialities with respect to other territorial processes; and (ii) the impact of tourism on local territorial systems and on their autonomy. What follows, then, is a simple model of tourist territorialization; 'simple' as it is but one methodological approach which allows us to analyse tourism as a particular geographical phenomenon which is necessarily interconnected with a variety of other geographical phenomena.

The model thus envisions a territorial system (S_1) related to other territorial systems in a multi scalar perspective (S_2, S_3, ... , S_n). This system is informed by several territorialization processes (T_1, T_2, ... , T_n), each promoting a specific vision of territorial organization (emergent, as it is, from a particular social and spatial project, that is, a particular idea on what that system is and what it should become). As we noted at the outset, it is precisely the dialectic between these different projects and their spatial processes that gives shape to the territorial system (Fig. 7.1; Minca and Draper, 1997).

The system, therefore, should be conceived of as a web of networks and nodes related to these diverse territorial projects. The landscape of the system, similarly, is made up of geographical iconographies associated with different territorial processes (or the interconnections between them). The system and its spatial organization are framed by conflicting and synergistic relationships; it is through the interpretation of these relationships that it is possible to identify the impulses (and thus the social actors behind them) that contribute to forge to the system.

The territorial project and the related spatial organization can be *endogenous* (therefore, made up of auto-centric strategies) or *exogenous* (made up of hetero-centric strategies). Each project can, however, express a different process depending on its intended scale; this means that tourist development can be 'planned' locally, nationally or internationally. This vertical relationship between the global and the local dimension is graphically illustrated by the arrows that represent the same process (like the one activated by tourist development) in its different strategies: exogenous (T_1, S_n for instance) versus endogenous (T_1, S_1).

Following this framework, tourist development can, therefore, be considered as an existing or potential process of territorialization (that is, as a strategy that influences the spatial evolution of the system itself). Tourist planning, therefore, represents a means to achieve a specific territorial configuration. The success of a tourist initiative, thus, depends on the relationships that tourist territorialities develop with other territorialities of the system. For this reason, the process of regionalization (as the reinforcement of the

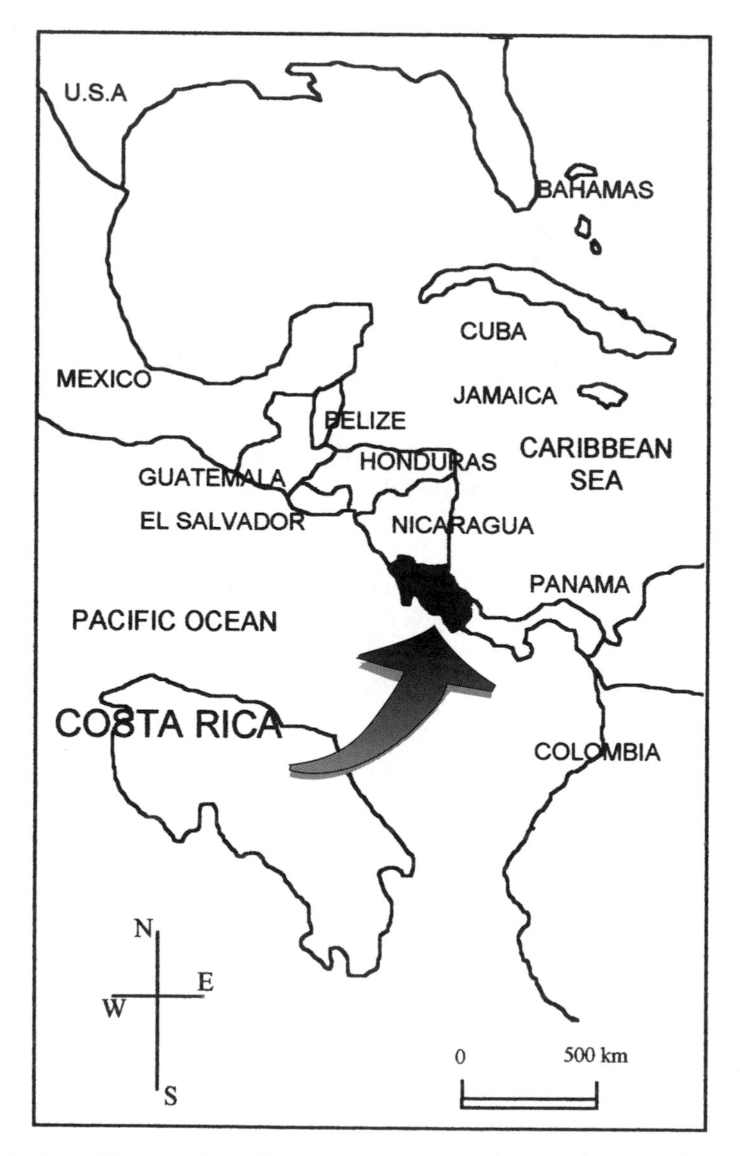

Fig. 7.2. Costa Rica and Central America. Source: Authors' elaboration from Baker (1994).

regional structure and identity) is strongly influenced by the contextual pressures of different social actors and their projects. Obstacles to a successful tourist re-territorialization might thus derive either from environmental obstacles or from the inertia of previous territorializations. Tourist territoriality, therefore, must be managed in order to avoid competition with other territorial processes (industrialization, environmental conservation, urbanization, agricultural development, etc.), as such competition might result in a progressive de-regionalization of the system, which also entails a progressive decline of the appeal of the destination.

On the other hand, when tourism is conceived as an explicitly synergistic strategy, it can represent a factor of regionalization, possibly even enhancing the autonomy of the system. This second hypothesis, though, is rather rare as it requires a very well-structured economic and social fabric; if this is the case, tourism can represent a significant and complementary source of income/employment. Yet, as noted above, it is the opposite case which is much more frequent: tourist development, being a strongly hetero-centric process, jeopardizes the regional/system structure.

In fact, traditional social hierarchies are often jeopardized by the cultural impact of the tourist economy; for this reason some regions prefer alternative development options. In most cases, however, the opportunities offered by the tourist market are too attractive for the local community (and, moreover, for regional and national governments) to pass up; these actors open their systems to the influences of the powerful tourist re-territorialization, often losing control of their very economies. This choice necessarily implies a progressive destructuration of the normative code of the system, particularly since the innovations introduced by tourism result in the consolidation of a new set of rules and spatial logics; rules, however, determined by exogenous projects.

This brief reflection warrants some final remarks. First of all, we note that the quality of the impact of tourist territorialization depends largely on the mechanisms of control that the system has activated with respect to external relationships. If the systemic 'code' is either weak or rigid (and thus poorly adapted to the absorption of innovation), the system is likely to undergo a destructuration. Otherwise, if the innovation (tourism) is properly metabolized thanks to the flexibility of the code, then tourist development can represent a substantial enrichment of the opportunities available to the system. We note, therefore, that when there exists a latent friction between tourist territorialities and other (existing) territorialities, the break-up of the system is quite likely in the long term. Irreversible tourist impacts (such as the transformation of natural resources/landscapes, the conversion of traditional activities, the consumption of strategic spaces) thus require a careful analysis to judge their consequences for the local community or region involved.

Development, Tourism and Sustainability in Costa Rica

Costa Rica, with its long history of struggle over the valorization of its natural resources represents, perhaps, an ideal case study for an examination of the role of tourist development processes in shaping the local territorial system. Following a 30-year-long period of large-scale natural resources exploitation, Costa Ricans realized the enormous value of their natural wealth in the early 1970s, starting the process which led to the protection of one-quarter of its

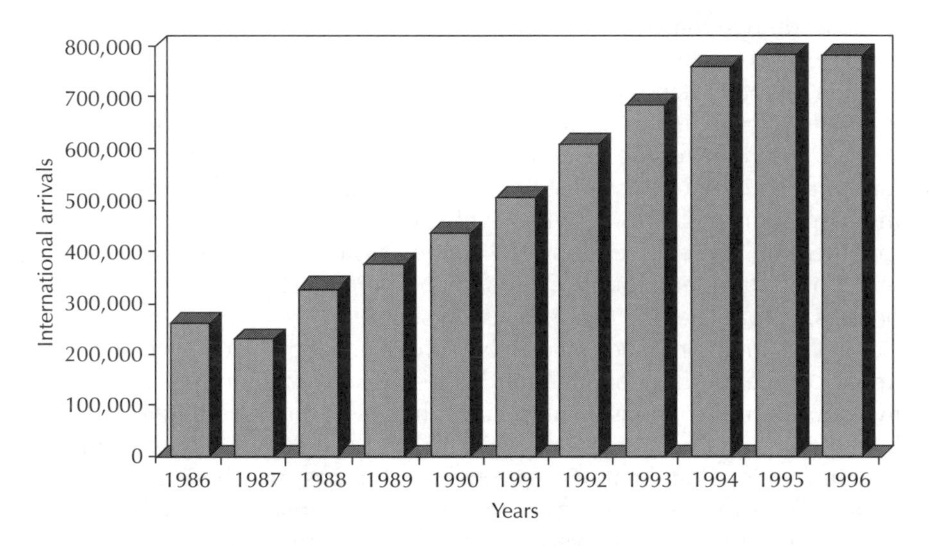

Fig. 7.3. International tourist arrivals in Costa Rica (1986–1996). Source: Authors' elaboration from ICT (1992), Market Data (1994), Universidad de Costa Rica and ICT Internet Site (www.tourism-costarica.com).

territory and to the award of a number of international conservation prizes. Commitment to conservation has brought a considerable amount of problems that Costa Rican governments through the years have not been entirely able to tackle. The creation of protected areas has, for example, allowed for the total depletion of natural areas lying outside of the preserves themselves at one of the fastest rates of deforestation in the world (Fig. 7.2; Baker, 1994).

With the growing importance of the ecological movement worldwide, Costa Rica foresaw a possibility to finance its commitment to conservation through some form of nature-based tourism and, around 1985, began to promote itself under the slogan 'Costa Rica es ... natural' (Champion, 1994). By 1989, tourism provided already 14.5% of the country's total foreign exchange, becoming the third largest source of foreign currency (after coffee and bananas), while in 1996 this percentage had risen to 23.4%. The number of tourists, in the period 1987–1997, grew at a remarkable annual average rate of about 14% (ICT, 1992, 1998; Market Data, 1994; Fig. 7.3).

The badly needed diversification of the economy seemed to be edging towards its realization: tourism was seen as a 'clean industry' and as such further encouraged. Striking a balance between the need for foreign currency and the commitment to nature conservation was no easy task, however. Although sustainability and ecotourism fast became the buzzwords in Costa Rica, the expectations invested in this road to development are not at all certain to be fulfilled; nor is the distribution of its benefits certain to be equitable (Linda, 1995).

In the early 1980s tourism emerged as a possible solution both to diversify the economy (thus mitigating the pressure on protected areas) as well as to support these areas' protected status. According to Budowski (1990), tourist interest was also boosted by the increasing numbers of scientists and researchers who began to converge upon Costa Rica since the 1960s. Researchers were followed by an increasing flow of nature-lovers, giving birth to what could be seen as a first phase of ecotourism development in Costa Rica, a stage which was characterized by a scarcity of capital, small-scale facilities and a more environmentally oriented approach. It was only in 1985 that the Government, operating through the ICT (Instituto Costarricense de Turismo), dedicated some resources to launch a campaign based on the natural beauty of the country and a programme of financial incentives aimed to encourage private investors.

It was only following this campaign that investments grew considerably, leading to a broad-based change in the nature of the infrastructures which expanded not only in number but also, and chiefly, in size. The ICT (1994), in fact, estimated that tourist offerings increased 16-fold in the period from 1987 to 1992. During the 1990–1994 legislature, a large-scale resort model of tourism development was promoted in order to position Costa Rica as a more traditional destination and to compete with the mass-tourism destinations of Mexico and the Caribbean. The North-Pacific coast has been particularly affected by this sudden shift from an environmentally friendly, élite form of tourism, to a low-budget type of mass tourism.

At this point in time, tourism had become the main source of foreign currency earnings having surpassed coffee in 1990 and bananas in 1993. The economy's heavy dependence on the exportation of coffee and bananas in the mid-1980s – when these exports accounted for 65% of total earnings – dropped to a mere 33% in 1996 (Chant, 1992; Champion, 1994; Market Data, 1994; ICT, 1998).

Although the boost in tourism revenues has been quite vital to the Costa Rican economy, the broad-based consequences are not yet entirely evident. It is as yet unclear how much of the returns of the massive foreign investments will stay inside the country, and how much will the local communities and the protected areas benefit from these returns.

In this chapter we shall focus our study upon the Osa Peninsula, located on the south-western coast of the country. This is the site of one of the last tropical rainforests in Central America, protected virtually in its entirety within the Osa Conservation Area. Its amazingly rich ecosystem provides an invaluable resource for an array of different activities from scientific research investigation to timber industry exploitation, through conservation, pharmaceutics and tourism. Its isolation from the rest of the country (300 km from San José) has saved it from exploitation – though also prevented its further development. The divide between conservation and development is thus particularly pressing in this delicate area.

Territorial Processes in the Osa Peninsula and the Corcovado National Park

The Corcovado National Park represents a particularly interesting case study for the analysis of the dilemmas and potential conflicts brought by Costa Rica's policies towards a more diversified economy. As outlined above, the park hosts one of the last and, according to Rachowiecki (1993), the best-preserved remaining tract of Pacific Coastal Rainforest in Central America. Located in the south-west region of the country, it is one of its wettest areas, with more than 6000 mm of rainfall annually, and is the richest in biodiversity: its forests contain 1000 species of trees, distributed with a density as high as 100–120 species per hectare in some areas (Watson and Divney, 1992), 140 species of mammals, 40 species of fish, 367 species of birds, 177

Fig. 7.4. Costa Rica and Osa Peninsula. Source: Authors' elaboration from Watson and Divney (1992).

Table 7.1. Conflicts and interactions in the Osa Peninsula.

Territorial processes	T_{1a}	T_{1b}	T_{1c}	T_{2a}	T_{2b}	$T_{1,a,int}$	$T_{2,a,int}$
T_{1a}	\	I,C	I	C	C	I	C
T_{1b}	I,C	\	I	I	C	I,C	I
T_{1c}	I	I	\	C	C	I	C
T_{2a}	C	I	C	\	I	C	I
T_{2b}	C	C	C	I	\	C	I
$T_{1,a,int}$	I	I,C	I	C	C	\	C
$T_{2,a,int}$	C	I	C	I	I	C	\

Source: Linda (1995).
T_{1a} = environmental organizations; T_{1b} = governmental institutions; T_{1c} = tourist investors; $T_{1,a,int}$ = international environmental organizations; T_{2a} = forestry industry; T_{2b} = agriculture and mining; $T_{2,a,int}$ = international forestry investors; I = interaction; C = conflict.

species of reptiles and amphibians and nearly 6000 species of insects. Some of these species, in fact, are threatened with extinction: most notably, is the jaguar, very rare in the rest of the country, as well as the scarlet macaw, whose population in Corcovado National Park is the most numerous in Costa Rica (Murillo, 1994; Fig. 7.4).

The framework for a systemic analysis of tourist territoriality and local development delineated in the introductory sections of the chapter could, perhaps, serve to analyse some aspects of the tension between the processes of conservation and development operating in this area. Tourism might induce an intricate network of relationships which could contribute to a destabilization of the territorial system involved and, consequently, to a loss of its autonomy. This is particularly crucial in the case of ecotourism, where development often implies inherent contradictions. Ecotourism, in fact, claims to support the environmental 'cause' and to bring development without detrimental impacts on the territory and on the (often) fragile ecosystems which are the very base of its prosperity. However, although as a principle it is easy to embrace, it is often utilized by commercial operators as but another marketing ploy to enlarge their market share (Wight, 1993; Cater and Lowman, 1994).

Following the framework of the territorialization model, the Corcovado National Park and the Osa Peninsula (which can be considered as its main area of influence) could be identified as a territorial system (S_1). This area will therefore represent the focus of the analysis and the portion of territory where the different processes of territorialization converge to give birth to a complex array of conflicts and interactions (Table 7.1).

As shown in Fig. 7.5 two main territorial processes can be identified in the analysed area: T_1, associated with conservation practices and T_2, which encompasses all projects linked to economic development strategies.

It is important to note that the agents grouped within each territorial process are not entirely coherent with the general strategy attributed to the mainstream 'ideology'. The forestry industry, for example, although clearly acting within the traditional economic scheme (T_2), has recently realized the strategic importance of embracing a 'more sustainable' approach (T_1) to its activities. As we noted above, the ecological awareness of Costa Ricans is very high and deeply rooted within the political culture of the country; it is not surprising that almost every organization operating in the Osa Peninsula claims to be highly committed to the cause of sustainable development.

The inclusion in the same process of territorialization (T_1) of the projects of the environmental organizations ($T_{1,a}$), of the governmental policies ($T_{1,b}$) and of the business strategies of tourist investors ($T_{1,c}$) could appear simplistic. However, as the system analysed is in a delicate phase of transition, the different processes do not have clear and fixed lines of conduct. The government, for example, notwithstanding its commitment to environmental issues, finds it difficult to balance the extreme need for development in the Osa Peninsula with the conservation policies. Similarly, while the contribution of the tourism sector to the environmental cause (through the protection of some areas, the pressure on governmental policies and the education of tourists) is considerable at the moment, considering the difficulty in control-

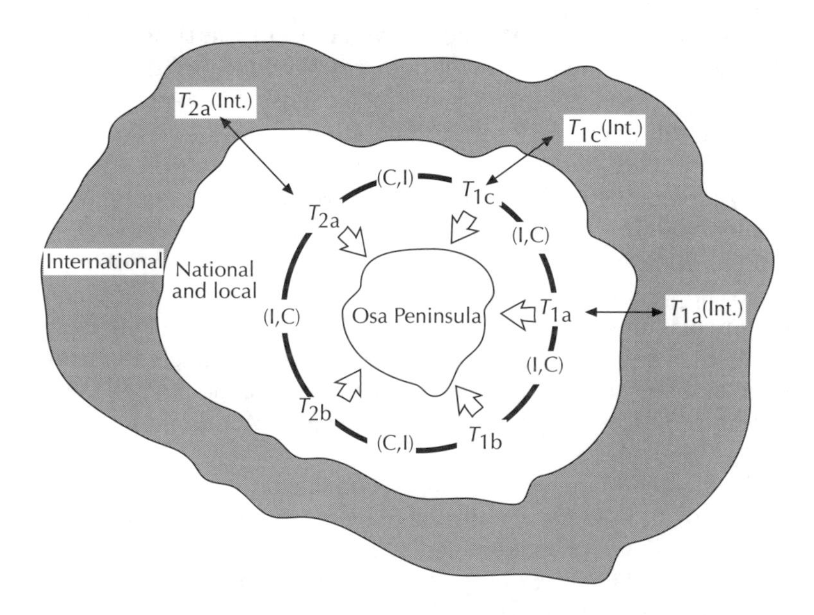

Fig. 7.5. Territoriality in Corcovado National Park and Osa Peninsula. T_{1a} = environmental organizations; T_{1b} = governmental institutions; T_{1c} = tourist investors; T_{2a} = forestry industry; T_{2b} = agriculture and mining; I = interaction; C = conflict. Source: Linda (1995).

ling tourism development (particularly due to the remoteness of the area and to the scarcity of funds), the potential for future negative impacts is high. Tourism development, in fact, is still at an embryonic stage at present, thus limiting its potential negative impacts on the territory.

Within the opposing territorial process (T_2) the action of the forestry industry is identified with $T_{2,a}$ whereas $T_{2,b}$ indicates other processes of traditional use of natural resources by the local communities (mainly agriculture, cattle farming and gold mining).

As stressed previously, the richness in biodiversity and the high degree of conservation have attracted a large number of environmental non-governmental organizations (NGOs) to the Osa Peninsula. These organizations act at the international, national and local levels. The predominant focus of the hetero-centric processes linked to the environmental ideology, through the projects of the *international organizations* ($T_{1,a,int}$) (mainly North American in origin), is that of fund-raising, with little involvement in practical activities 'on the ground'. The European Union, with the Project of Integrated Rural Development for Osa and Golfito, is also particularly involved in the area.

A significant event marking the strong role currently held by environmental organizations ($T_{1,a}$) in the area is the intervention of Greenpeace in the dispute with Stone Forest Inc., the main exogenous actor of the forestry industry ($T_{2,a,int}$). In August 1994, after the positive result of an Environmental Impact Assessment process (F. Bermúdez, Costa Rica, 1994, personal communication), Stone Forest obtained government permission to build a chip mill and a dock in the northern area of Golfo Dulce (Punta Estrella; Fig. 7.6). Environmental activists claimed that the delicate ecology of the Golfo Dulce would have been threatened by the operations of the mill (A. León, Costa Rica, 1994, personal communication). Due to the public outcry, as well as the actions of the local and national NGOs, the project was reviewed and the permission was withheld in November 1994 (Harris, 1994).

At the *national and local level* $T_{1,a}$ is expressed by the action of different non-profit organizations; amongst the most active are the Neotropic Foundation (NF) and the INBio (Instituto Nacional de Biodiversidad). The NF was founded in 1985 and its main objectives on the Osa Peninsula are, to quote Cabarle *et al.* (1992), 'to develop and demonstrate natural forest management, sustainable agriculture, ecotourism, and biodiversity technologies which are economically productive and contribute towards the maintenance of forest cover'. The NF project for the Peninsula (BOSCOSA) is confined by legal constraint to just a part of the buffer areas surrounding the park. Only 13,000 of the nearly 77,000 ha of the Golfo Dulce Forest Reserve are thus part of BOSCOSA. In this area, the project exists as a collaborative endeavour with the Ministry of Natural Resources (MIRENEM) and through its various line agencies which have juridical responsibilities over the different areas according to their status of protection (NF, 1989). The main objective of the ecotourism programme, activated in 1991, has been to involve and train the

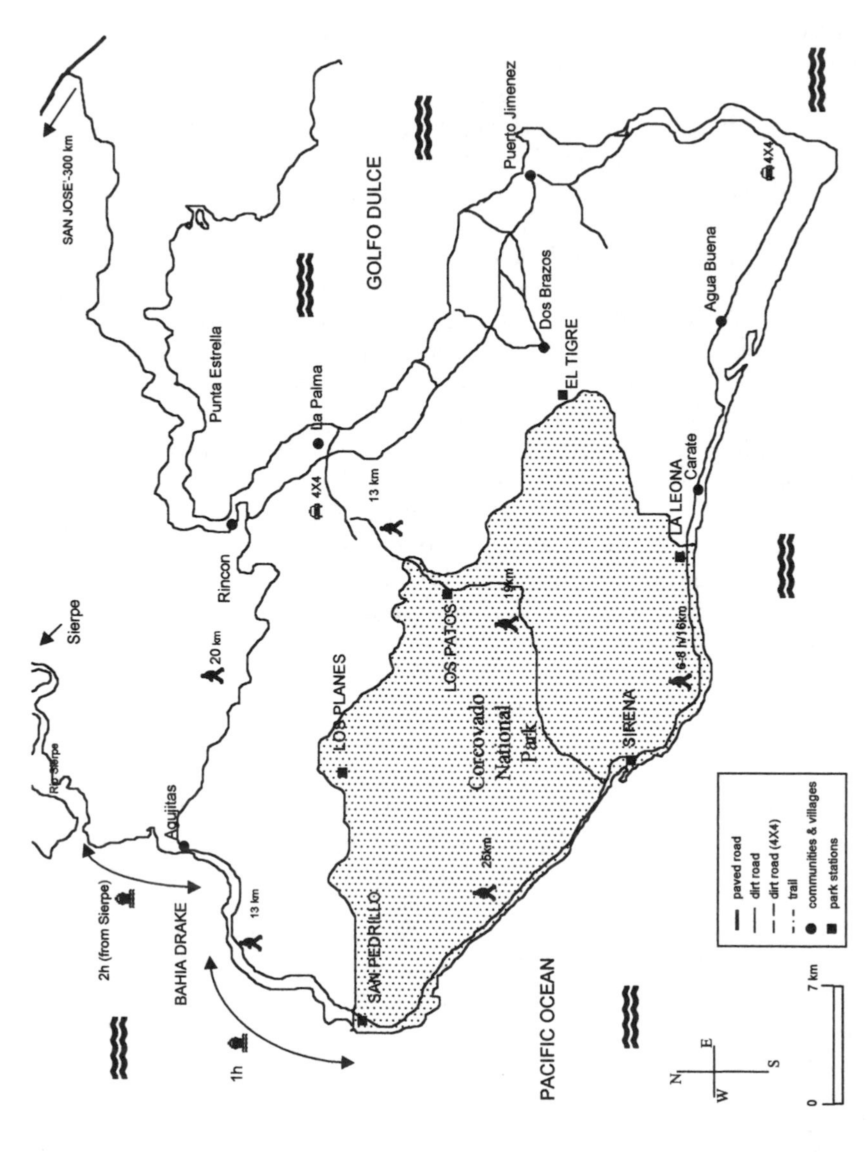

Fig. 7.6. The Osa Peninsula and Corcovado National Park. Source: Linda (1997).

local communities in order to make them aware of the importance of natural resources (Jiménez, Costa Rica, 1994, personal communication; R. Vargas, Costa Rica, 1994, personal communication).

The INBio, founded in 1989 and partly funded by the Government, is similarly important in our analysis as it is a clear example of the commitment of the Costa Rican authorities to the environmental cause. The most challenging project currently carried out by the INBio is the National Biodiversity Inventory, whose aim is to collect and rationalize data on all the species existing in Costa Rica. INBio is active with four biodiversity field offices employing local inhabitants in the area of the Osa Peninsula and Golfito (Sittenfield and Villers, 1993, 1994).

The government with its complex bureaucratic network is indicated in the model by $T_{1,b}$ and, as the direct administrator of the entirety of the protected areas of the Osa Peninsula (nearly 80% of its total surface), is involved in the main territorial process of the area. Corcovado National Park is fully owned by the national government, which is also responsible for the administration of the other protected areas of the Osa Peninsula. Yet the heterogeneity of interests, coupled with the pressures deriving from the conflictual demands of development and conservation, have rendered the government's task far from simple (Linda, 1997).

It is the territorial processes linked to tourism development ($T_{1,c}$) that are, however, the main focus of our analysis. As argued above, the international tourist investors (chiefly those from North America) possess a fundamental and at the same time contradictory role both at the national and at the local levels. In the case of the Osa Peninsula, due to its characteristic underdevelopment and isolation, the paradox is even clearer. The international tourist investors hold a formidable advantage in reaching the market; they are better organized both in terms of transportation to the Peninsula and in terms of range and quality of the activities offered; they can also offer higher standards and afford the expensive disposal techniques necessary in the areas where the most basic of public services (water, sewage and electricity) are poor if not entirely non-existent (Minca and Linda, 1999).

It becomes quite apparent why foreign tourist investors have become concentrated in the least accessible and developed area of Bahia Drake where the Corcovado N.P. and the Isla del Cano are located. These are the main attractions of the area and can both be reached in one hour using the same powerful boats which bring tourists from Sierpe. This area is mainly dedicated to the most affluent segments of tourist demand – high priced and highly specialized hotels and activities. The exogenous territorial pressure of the international tourist investors ($T_{1,c,int}$) on the system (S_I) is minor given the limited rate of development due to the isolation of the area from the rest of the country. The effect of the local investors ($T_{1,c}$) is even more limited as they can host lower numbers of visitors even if, paradoxically, their operations are very often more polluting that the ones of foreign investors, that can afford expensive sewage control systems.

The local communities of the Osa Peninsula, encouraged by the active support of the NGOs ($T_{1,a}$) and the government agencies ($T_{1,b}$), have great expectations for the future of ecotourism. Owing to the training and to the fund-raising assistance provided by the organizations mentioned above, they have managed to set up a number of small enterprises in the last 6–7 years. As the level of services offered is quite limited when compared to the hotels run by foreign investors, the only segment of the market that the local businesses can target is that of backpackers or scientific researchers.

Passing on to examine the effects of the *economic development territorial process* (T_2) the activities of the *forestry industry* ($T_{2,a,int}$) will be analysed first. The main actor within this process is the Stone Forest Inc., a leading multinational of the paper industry whose leaflets assert a clear orientation to sustainable development: their aim is to re-forest 60,000 acres of despoiled land in the Southern areas of Costa Rica, employing 1800 local people and renting their land, thus benefiting the local economy. In 1994, the above project reached completion, with investments for $16 million and the creation of 1300 jobs.

Local communities ($T_{2,b}$) have always been dedicated to subsistence farming and gold-mining as well as forestry activities. NGOs and government agencies have implemented a variety of aid programmes in order to improve their cultivation techniques and to render their activities more sustainable. Yet the *campesinos* and the *oreros*, after the expropriations necessary for the creation of protected areas, do not trust the actors of the environmental territorial process. After a long period of non-intervention, the government has rapidly increased its control over the area through strict zoning and a new legal framework for forest management. These actions were far from popular: while previously local people could settle and work on any land which was not already occupied by somebody else, with the new regulations, settlement necessitated a long and complicated process. Local people have found it easier to sell timber or even land to foreign investors with consequent social and economic impacts (Cabarle *et al.*, 1992). In still other cases, farmers, loggers and miners have entered the protected areas illegally, causing even worse damage to the forest.

The Evolution of Territorialization in the Osa Peninsula

To fully understand the intricate present-day configuration of the territorial structure of this system, the past evolution of its territoriality must be examined thoroughly, as its role on the current territorial instability of the system as well as on the main conflicts that are contributing to it is considerable. The territorial evolution of the Osa Peninsula can be divided into two main phases, with the foundation of Corcovado National Park in 1975 seen as the watershed event. In the first period, from 1848 to 1975, developmental ideologies (I_2) clearly prevailed and it was only in 1975 that the conservationist forces (T_1) began the processes that would revolutionize the structure of the system in only 30 years (Table 7.2).

The history of territoriality in the Osa Peninsula points to the crucial role of the government ($T_{1,b}$), as well as its openness towards the hetero-centric processes of the multinational companies ($T_{2,a,int}$) that, over time, developed increasing interest in the natural resources of the area. The evolution also demonstrates the changing importance of territorial processes originated by the local population ($T_{2,b}$); once the only processes in the system, though fading in importance through the various interventions of the government. Until the 1920s, the area of the Peninsula was completely underdeveloped and scarcely populated. Local inhabitants, through slash and burn techniques, contributed heavily to the deforestation of the area. In the 1930s, the United Fruit Company began its activities in the area, settling its banana plantations in the north of the Peninsula. This was to be the first of a series of major events which contributed to the colonization of the Peninsula.

Yet another major event impacted the socio-economic development of Osa: in 1938, gold was discovered in this area, a discovery that triggered yet another phase of colonization of the Peninsula (Murillo, 1994). Several concessions for the exploitation of gold resources were ceded to national and international companies, as well as to individuals (Castillo *et al.*, 1993). However, the gold was not as abundant as it appeared and miners began to settle and to cultivate extensive areas, a factor which contributed considerably to the deforestation process. It was in the late 1950s that the United Fruit Company began its decline, though still owning 30% of the total surface of the Peninsula. In 1959, these properties were to be sold to the Osa Productos Forestales, a multinational of the forestry industry which started to exploit the natural resources of the Peninsula. Again, the forests of the central areas were left untouched, as the high heterogeneity of plant species rendered the exploitation not viable economically. The powerful multinational also obtained the concession from the government to exploit a further 40,000 ha. This action was to begin a long-standing conflict with the *campesinos* who were already working on that land.

All was sacrificed in the name of developmental ideology (I_2) during this period. The tensions between Osa Productos Forestales and the *campesinos* increased to such a level, that its very existence – and its real contribution to the economy – began to be questioned at the governmental level. In 1975 Corcovado National Park was founded as part of the national policies of conservation: local residents and *oreros* were relocated and portions of the logging company's concessions withheld.

After centuries of prevalence of the developmental forces, the first acts of the territorialization processes linked to the conservational ideology encountered a stiff opposition. Local people, in particular, could not accept the renunciation of the resources they had always exploited. The conflicts between the *campesinos* ($T_{2,b}$) and the multinational ($T_{2,a}$) were also

Table 7.2. Territorial evolution of the Osa Peninsula.

Period		Territorial elements	Conflicts and interactions
1848–1930	T_{2b}	• Colonization (agriculture and cattle farming)	• T_{2b} dominant
1931–1975	T_{2b}	• Increasing expansion and deforestation	• T_{2a} dominant
		• Gold mining	• First C between T_{2a} and T_{2b}
	T_{2a}	• First investments of theUnited Fruit Company and further expansion	
		• First investments of the Osa Productos Forestales	
	T_{1b}	• Extension of the Pan-American Highway	
1976–1988	T_{1a}	• First actions of the environmental organizations and foundation of BOSCOSA	• T_{1a} and T_{1b} dominant
	T_{1b}	• Creation of Corcovado National Park, Golfo Dulce Forest Reserve and other PAs and consequent land expropriations	• Increasing C between T_{2a} and T_{2b}
		• Expulsion of the *oreros* from the Park	• C between T_{1a} and T_{1b} and T_{2a} and T_{2b}
	T_{1c}	• First foreign investors (*Bahía Drake*)	
	T_{2a}	• United Fruit Company and Osa Productos Forestales cease their activities	• Early actions of T_{1c}
	T_{2b}	• Increasing expansion of agriculture and clandestine mining activity	• Early evidence of I between T_{1a} and T_{1b}
1989–	T_{1a}	• Further expansion of the projects	• T_{1a} and T_{1b} dominant
	T_{1b}	• Concessions to T_{2a}	• C between T_{1a} and T_{1b} and T_{2a} and T_{2b}
	T_{1c}	• Further development	
	T_{2a}	• Growth in forestry activities in the buffer areas	• Open I between T_{1a} and T_{1b} and first evidence of I between T_{1a}/T_{1b} and T_{1c}/T_{2b}

Source: Linda (1995).

becoming increasingly harsh; a situation that led to the definitive expulsion of the logging company from the Osa in 1979 (Murillo, 1994).

The tourist territorialization processes ($T_{1,c}$) have begun only recently, largely with the encouragement of the government which realized the potential contribution of the tourist industry to the solution of the conflicts between conservation and development. The first tourism infrastructures were actually realized in the 1930s in Puerto Jimenez, when gold was discovered and the activities of the United Fruit Company began, but when business declined, most had to close down. The improved accessibility, together with the growing interest in the protected areas, has significantly increased the flows of tourists, and the accommodations of Puerto Jimenez began to host the first 'ecotourists'. In the early 1970s, the Osa Productos Forestales, given the opposition encountered by its logging activities, tried to persuade the government to back the project for a big tourist resort in Bahia Rincon (northeastern side of the Peninsula). In 1977, the multinational was expelled from the Peninsula and the project was never begun. The first investments linked to 'ecotourism' date back to the beginning of the 1990s, when some American investors began their activities in the area of Bahia Drake. Although growth has been very slow, the latest years have witnessed a growth of infrastructures, inspired by increasing numbers of visitors to the Park.

Ecotourism, Sustainability and Territoriality in the Osa Peninsula: A Complex Network of Conflicts and Interactions

The process of territorialization ($T_{1,b}$), which represents the policies of the government, can be considered as the most crucial at the moment. As the national government is responsible for the conservation of 80% of the total surface of the system (S_1), it appears to be the principal shaper of the environmental ideology. However, as we have pointed out above, it has had a very difficult task in balancing conservation and development interests and consequently has had a very ambiguous role in dealing with the opposing ideology. This weakness can be considered the main source of the conflicts that the process ($T_{1,b}$) has to overcome in the realization of its projects upon the territory.

Although the government's relationships with the other territorialization processes are mainly collaborative, some conflicts do exist. The main conflicts, both in the past as well as at present, originate from the relationship with the local population ($T_{2,b}$) and, in particular, with the *oreros* who, when the Park was created in 1975, were banned from it due to their supposedly detrimental impacts on the environment. The relationship with the farmers, although improving, is still problematic. The origin of these conflicts can be found in the expropriations forced by the government and in the restrictions on the use of land introduced during its policy of conservation. As a consequence of the conservationist policies of the government in the past three decades, local people have become similarly suspicious of the projects of the environmental NGOs ($T_{1,a}$) which are perceived as yet another danger.

The interaction between the government ($T_{1,b}$) and the conservationist groups ($T_{1,a}$) would seem particularly clear as many governmental programmes are implemented by the Neotropic Foundation and as there is a contractual relation with the INBio. However, the favourable attitude of the authorities towards Stone Forest Inc. ($T_{2,a}$) has worsened the relationship. In addition, there is significant competition in fund-raising activities between the two.

The relationship between environmental NGOs ($T_{1,a}$) and the Stone Forest ($T_{2,a}$) is openly conflictual (J.R. Vargas, Costa Rica, 1994, personal communication). The latter has been accused of polluting the environment as well as excluding local participation through buying and renting their land. Some locals, in fact, have referred to Stone Forest as the 'conquistadores' (M. Villalobos, Costa Rica, 1994, personal communication); others, however, support them because they are creating job opportunities, a very scarce resource in the Peninsula.

This, then, is the complex and unstable background in which tourism development ($T_{1,c}$) has started to expand, slowly but increasingly, in the last 8 years. As we have noted previously, the territorial autonomy of the system is already quite low due to the complex network of conflicts around the territorial use. We expect that tourism, introducing an external and destabilizing process of territorialization, will probably contribute to the de-regionalization which currently characterizes the Osa Peninsula.

Despite the fact that the relationship with the main actors is, at present, collaborative (as the data collected through the interviews have indicated), there are already signs for the development of potential further conflicts. Both the government ($T_{1,b}$) as well as the conservationist groups ($T_{1,a}$) have encouraged local people to set up their own ecotourism businesses. At the same time, government policies are facilitating foreign investment. Environmental NGOs ($T_{1,a}$) have tried to promote the development of a form of ecotourism which respects the principles of sustainability (that is, when such tourism can be integrated, in a synergistic fashion, within the territorial system's normative code). Yet the foreign investors supported by the government have only created menial jobs for the local population which, if able to set up independent businesses, has been forced to target the lower segments of the tourist demand. An intrinsic contradiction thus exists; a contradiction likely to result in conflicts over the use of the territory.

The relationship of tourist territorialization with local farmers ($T_{2,b}$) is almost non-existent and constitutes another possible source of conflicts. To preserve the natural beauty of the Osa Peninsula, local people have been deprived of their base source of income – and they have received very little in exchange. Environmental NGOs are currently trying, through training programmes, to improve farming techniques to enhance the quality of the crops which could then be sold to tourism businesses. This strategy would hopefully solve the problems of isolation and distance from the market that afflict farmers. Local people could consequently benefit from tourism development ($T_{1,c}$) which could contribute to the stabilization of the system. But, as outlined above, NGOs have still to gain the confidence of the farmers who are very sceptical about conservationist policies.

The relationships between local investors in the tourism sector and investors from outside the area are, currently, interactive. Local businessmen do not see outside investors as a threat as they serve a very different segment of the demand. Some other local people even see them as benefactors: in the area of Carate, for example, funding was obtained for the construction of a school and in the area of Bahia Drake foreign investors have also provided a special rate (a quarter of the price charged to tourists) for the boat service to the nearest town of Palmar Norte, where locals can find their basic necessities. This, however, has been the limited extent of the contact with the locals; the tourists have certainly not ventured from the secure areas of their hotels into the nearby community which lies only within a walking distance of 10 min.

Conclusions

The application of the systemic model of tourist territoriality to the Osa Peninsula can permit an initial understanding of the evolution of the different territorial processes and their inter-relationships. In particular, it is clear that if tourism is to provide an alternative to the traditional non-sustainable exploitation of natural resources and a means towards the realization of territorial autonomy, the ground for an interactive relationship with the existing processes of territorialization must be prepared.

In less than 20 years, as seen in the analysis of the historical development of the territorialization of the Osa Peninsula, conservationist ideology has radically transformed the socio-economic structure and the territorial organization of the area, leaving its inhabitants even poorer than before, while transforming the Osa Peninsula into the 'ultimate' attraction for ecotourists. The number of businesses involved in the ecotourism sector is in fact rising at a sharp rate. Economic sustainability should be taken into account in any ecotourist development: this would bring benefits to local people in order to increase their autonomy and to improve their livelihoods. At present, as the field study has highlighted, the tourism sector provides local people only with low-paid jobs.

The conservation NGOs are contributing greatly towards the involvement of local people in the tourism sector, as well as towards the improvement of their agricultural techniques through the programmes of training and environmental education described previously. Yet while they have succeeded in establishing a good relationship with the potential tourist investors, they still have to gain the confidence of the small farmers. It is difficult to convince the locals not to use certain noxious chemicals or to start the cultivation of alternative crops when these latter are involved in a daily struggle to feed their families. After the expropriations made in the name of the conservationist ideology, it is difficult to enforce cooperation with the environmental 'cause'.

The government, in its difficult role of balancing conservation and development, has found a possible solution in ecotourism development. Its relaxed attitude towards foreign investors and logging companies, however, is not encouraging sustainability. Such a dilemma is, perhaps, unsurprising, considering the government has a chronic need for foreign currency and the high levels of public debt?

Considering the above, it is quite evident that tourism development, in its current evolution, does not contribute towards the strengthening of the territorial system in question. This system's autonomy is, in fact, already quite limited, reflecting the consequences of the contradictory past interventions of public authorities. Unless a more sustainable planning approach to tourist development is implemented, the risk that it contributes to a loss of autonomy and control by the local community (and, consequently, to the process of de-regionalization) is quite high.

The application of the model has thus aided in the analysis of the actual territorial organization of the system in question, allowing a deeper understanding of the processes that have led to the current structure of the territory. An improved comprehension of the complex web of relationships within the system could prove to be a very important base for better planning of future development. Further research could focus more concretely upon local patterns of development, relying upon specific field surveys aimed to address key issues such as local involvement (qualitative and quantitative), the success of local tourism investors and actual improvement in sustainability brought by environmental NGOs' programmes.

References

Baker, C.P. (1994) *Costa Rica Handbook.* Moon Publications Inc., Chico, California.

Budowski, T. (1990) Ecoturismo a la Tica. In: *Hacia una Centroamerica Verte: Seis Casos de Conservacion Integrada.* DEI, San José, Costa Rica, pp. 73–89.

Cabarle, B., Bauer, J., Palmer, P. and Symington, M. (1992) Boscosa: The Program for Forest Management and Conservation in the Osa Peninsula, Costa Rica. Project evaluation report. Fundación Neotrópica, San José, Costa Rica.

Castillo, E.A., Quesada, S.E.C., Camacho, J.L. and Masis, A. (1993) Recomendaciones para un Ordinamiento Territorial en la Peninsula de Osa, Costa Rica. Programa Boscosa, Area de Ordinamiento Territorial. Fundación Neotrópica, San José, Costa Rica.

Cater, E. and Lowman, G. (1994) *Ecotourism: a Sustainable Option?* John Wiley & Sons Ltd, Chichester.
Champion, J.W.S. (1994) Ecotourism in Costa Rica: an investigation into community involvement, and its benefits and costs – with case studies. Submitted in partial fulfilment of the requirements for the award of the degree in Master of Science in Tourism, Food and Hospitality Management. Sheffield Hallam University, Sheffield.
Chant, S. (1992) Tourism in Latin America: perspectives from Mexico and Costa Rica. In: Harrison, D. (ed.) *Tourism and Less Developed Countries*. Belhaven Press, London, pp. 85–101.
Freeman, G.M. (1992) *Stone Container and the Miracle Tree*. A presentation by Gerald M. Freeman, President and Chief Operating Officer, Stone Forestal Industries, Inc., 24 September (1992) (brochure), Kansas City, Missouri.
Haber, H. (1993) *Costa Rica*. Guide Apa, Italian Edition of Insight Guides. ZanfiEditori, Modena, Italy.
Harris, B. (1994) Ston OKs Dock Site Relocation. *The Tico Times*, 18 November, pp. 1, 5.
ICT (Instituto Costarricense de Turismo) (1992) *Costa Rica. Tourism Statistics in Brief 1992.*
ICT (Instituto Costarricense de Turismo) (1998) Internet site: www.tourism-costarica.com.
ICT (Instituto Costarricense de Turismo), and European Community (1994) *Plan Estratégico de Desarollo Turístico Sustentable de Costa Rica 1993–1998.* ICT/ECC, San José, Costa Rica.
Linda, M. (1995) *Un Modello di Sviluppo Ecoturistico: il Caso del Parco Nazionale Corcovado.* Tesi di Laurea in Geografia Applicata. Facoltà di Economia, Università degli Studi di Trieste, A.A. 1994–95, Trieste, Italy.
Linda, M. (1997) Ecoturismo e Sviluppo Sostenibile in Costa Rica: Il Caso del Parco Nazionale Corcovado. *L'Universo* 2.
Market Data (1994) *Costa Rica: Datos e Indicatores Básicos. Costa Rica at a Glance.* INICEM, San José, Costa Rica.
Minca, C. and Draper, D. (1997) Territory and tourism: the case of Banff National Park. *Tijdschrift voor economische en sociale geografie* 87, 99–112.
Minca, C. and Linda, M. (1999) La natura tra parentesi 2: Il caso del Parco Nazionale del Corcovado (Costa Rica). *Turistica* 7, 51–74.
Murillo, K. (1994) *Perfil de Planificación o Estrategia de Formulación Operativa.* Metodologia de información (base: revisión cartográfica y bibliográfica). Area Protegida, Parque Nacional Corcovado, enmarcado dentro del Area de Conservación y de Desarollo Sostenible de Osa Junio 1994.
NF (Neotropic Foundation) (1989) *Osa 2000. Biological Conservation and Community Development for Corcovado National Park and Buffer Zone on the Osa Peninsula, Costa Rica.* Fundación Neotrópica, San José, Costa Rica.
Rachowiecki, R. (1993) *Costa Rica*. E.D.T., Torino, Italy.
Sittenfeld, A. and Villers, R. (1993) Exploring and preserving biodiversity in the tropics: the Costa Rican case. *Current Opinion in Biotechnology* 4, 280–285.
Sittenfeld, A. and Villers, R. (1994) Costa Rica's INBio: collaborative biodiversity research agreements with the pharmaceutical industry. In: Meffe, G.K. and Carrol, C.R. (eds) *Principles of Conservation Biology*. Sinauer Associates, Inc., Sunderland, MA, pp. 500–504.

Turco, A. (1984) *Regione e Regionalizzazione.* Angeli, Milano, Italy.
Turco, A. (1988) *Verso una Teoria Geografica della Complessità.* Angeli, Milano, Italy.
Vallega, A. (1982) *Compendio di Geografia Regionale.* Mursia, Milano, Italy.
Vallega, A. (1995) *La Regione, Sistema Territoriale Sostenibile.* Mursia, Milano, Italy.
Watson, V. and Divney, T. (1992) Informe Sobre el Mapa de Ordinamiento Territorial del Area de Conservacion Osa. Preparado para Ministerio de Recursos Naturales, Energia y Minas (MIRENEM). Colaboradores: Danilo Godoy y Jorge Jiménez. Centro Cientifico Tropical, San José, Costa Rica.
Wight, P. (1993) Ecotourism: ethics or eco-sell? *Journal of Travel Research* 21, 2–9.

8

Ecotourism in Tropical Rainforests: an Environmental Management Option for Threatened Resources?

Simon Evans

Introduction

The erosion of forest resources has been an issue of concern for many decades, particularly in relation to diminishing reserves of tropical primary forest. As environmental awareness has grown across society as a whole, so too have demands to visit such natural environments. Whilst environmental management techniques have been introduced in many parts of the world to offset the problems caused by over-visitation, in many tropical areas it is tourism itself which has been identified as an environmental management policy in its own right. Ecotourists seeking natural, unspoiled environments value rainforests for their authenticity and unfamiliarity, particularly in cases where their own local forest resources have been transformed by permanent artificial surfacing and fencing construction. In turn, the receipts derived from those tourists can be diverted back into the management of the resource, providing an incentive for host nations to protect rather than exploit their natural resource base.

This chapter considers the reality of ecotourism as a tool of environmental management and questions the extent to which authenticity is being compromised in the long term by environmental control requirements and visitor expectations. It proceeds to argue that, in the name of sustainable development, once sufficient visitation has been attracted to an area in order to provide financial support for conservation then, just as in many developed nations, additional forms of environmental control will inevitably follow. Furthermore, these practices may prove similar to the artificial constructions experienced in the visitors' home localities. The consequences of this may be an intensification of visitor numbers which can be accommodated at the ecotourism destination, more akin to mass tourism, and a continual process by which more adventurous travellers will seek new, unspoiled environments to colonize. This cycle may ultimately lead to loss of distinctiveness, the emergence of stereotypical developments on an international scale and, in certain instances, cultural dilution or manipulation to satisfy perceived tourism demands. The roots of this process can perhaps be traced to a deforestation–protection–tourism–development continuum which can develop around the theme of ecotourism.

Setting the Scene

The employment of forests for recreational and tourism purposes has long been acknowledged as an important feature of forest management in developed nations, a pressure which shows little sign of abating (Benson and Willis, 1991). The compatibility of access with timber production, in many situations the primary function of forest resources, has been widely recognized (Irving, 1987), yet the combination of access and conservation-related motivations has frequently proved more contentious and complex (Evans, 1998). Indeed, this relationship can be considered somewhat paradoxical, with the ecological and aesthetic attributes of forested environments generating visitor demands, yet the effects of that visitation, if unregulated, ultimately degrading the resource base and inhibiting future access demands. In the context of sustainable development (WCED, 1987), this scenario raises some important questions, fuelling calls for the increased regulation and protection of scarce forest resources in order to prevent irreversible losses.

In the developed world in particular, a range of resource and visitor management techniques have been formulated, tested and implemented in order to accommodate conservation objectives into multipurpose management strategies. These methods range in prescription from mildly coercive approaches such as education and interpretation to the imposition of officially sanctioned access routes via permanent surfacing and fencing construction, through to the prohibition of unauthorized access to zones of particular sensitivity (Cooper *et al.*, 1998: 476). In a UK context, evidence would suggest that some visitors may be deterred from visiting forests which utilize artificial materials as part of their management regime, considering these to be inconsistent with the rural character they demand as part of their forest experience (Evans, 1992). Furthermore, issues of vehicular access, traffic congestion and the presence of large visitor numbers may be viewed as a form of ephemeral urbanization, once more detracting from a site's perceived quality (Evans, 1992).

Whilst this situation differs significantly from conditions experienced in, and forces exerted upon, forest resources across much of the developing world, certain parallels nevertheless exist. What is more, these similarities are becoming increasingly evident as new tourism demands have emerged in tropical regions previously considered to be remote and largely inaccessible (Mowforth and Munt, 1998). While this geographical expansion of tourism may reflect technological advances and increased economic freedom, it also has its origins firmly rooted in a new environmental paradigm which has developed around the theme of sustainable development, an issue which is considered in detail within the introduction of this book.

Tropical Forest Resources – an Overview

Within the debates surrounding sustainability, the issue of forest conservation and expansion has occupied considerable space in the supporting literature. Despite their coverage of only 2% of the Earth's surface or 6% of its land mass (Rainforest Action Network, 1997), tropical rainforests are acknowledged to contain over 50% of the global stock of biological species (Madeley and Warnock, 1995). Such richness in biodiversity conveys a range of benefits upon both local populations and the wider global community. Although of obvious importance, it is not merely the plant and animal life alone which represents the total value of the resource, but the multiple roles and interrelationships which exist between the forest itself and a host of other scientific, economic, cultural and physical processes associated with its use. The inventory of genetic, species and ecosystem diversity contained within tropical forests support a variety of vital life support systems (Phillips *et al.*, 1994), including valuable medicinal properties and staple foodstuffs. This is regarded to be a resource which, once depleted, can never be fully replicated (Gomez-Pompa *et al.*, 1972), thereby affecting the needs and livelihoods of future generations implicit in the equitable principles underpinning sustainable development. Furthermore, threats of irreversible deforestation in the tropics may ultimately have severe global repercussions in respect to their role as climatic regulators (Brown *et al.*, 1993).

The capacity of trees in storing damaging greenhouse gases (Ciesla, 1995) has been a major issue in raising public awareness as to the prudence of forest retention. Arguments persist, however, that this mindset of environmental concern remains more prevalent in the developed world than elsewhere and that pressures exerted by dominant cultures of the northern environmental lobby may conflict with the everyday realities faced by indigenous communities. Indeed, a feature of the environmental concern attributed to indigenous cultures surrounds not just resource exploitation by outsiders but resistance to other impositions upon use emanating from external sources (Vivian, 1992).

As the vast majority of tropical regions fall under the direct authority of developing nation governments, the coincidence of forest resources with populations characterized as rural poor is largely unquestioned. The value of forest resources to such communities has been well documented (Arnold, 1992), with the term access used to describe much more than merely recreational or tourism opportunity as it has become associated primarily with in the context of the developed world. In this case, many of the basic subsistence needs of local communities have traditionally been satisfied within the precinct of the forest. Demands for firewood, fuel and construction materials, for food, fodder and agricultural space, have all been cited as essential needs facing rural populations in their use of local forest resources (Arnold, 1992).

Previous strategies aimed at the protection of forest resources have not always reflected such cultural and social realities, however, and the displacement of indigenous peoples from traditional territories has not been an uncommon phenomenon. This issue will be pursued in greater depth within the discussions concerning protected area designation. Increased calls for the protection of tropical forest resources reflect this multiple value base and the damaging effects associated with its destruction, which has been occurring for many years. Whilst there is little dissension surrounding the fact that tropical forests continue to be degraded and destroyed, the rate at which they are being exhausted and indeed the main agents in that process, are to some extent disputed through much of the existing literature.

Levels and Causes of Tropical Deforestation

If the reasons for tropical deforestation evade consensus acceptance, experts remain united in their assessment that exploitation is happening at alarming rates, albeit with significant regional variations (Pearce, 1998). Estimates that a mere one-fifth of the world's original old growth forests remain in large, continuous natural ecosystems (World Resources Institute, 1997), although damning in their own right, may not fully articulate the scale of the problem as it persists to the present day. Myers (1989) claimed that at current levels of loss, around 50,000 species per year were being driven to extinction every year. This problem has been exacerbated by the fragmentation of forest resources into smaller pockets often lacking spatial links to other similar areas (Skole and Tucker, 1993). Peters *et al.* (1989) present a case study example provided by research undertaken in the Amazonian rainforests of Peru to question the economic wisdom behind this approach. The authors claimed whilst clearfelling for timber extraction generated an approximate return of US$1000 per hectare in addition to a further US$148 per annum for subsequent cattle pasture, if sustainably managed, this figure would be nearer to US$7000 per year.

Certain sources cite the activities of commercial logging operators as the main agents of deforestation (Repetto, 1990; Winterbottom, 1990), particularly in areas where the conversion of timber resources into working capital enable governments to stimulate and sustain industrialization programmes. Additional effects associated with logging activities surround the infrastructure required for harvesting and extraction, with Myers (1989) estimating that for every tree removed for timber in Zaire, a further 25 are lost due to clearance for road access. In some areas of South-east Asia considerable pressures are exerted upon forest resources as a consequence of oil exploration (Rainforest Action Network, 1999), whilst certain areas of central America suffer from clearances to accommodate cattle ranching. One consequence of such deforestation is the utilization of formerly forested land for agricultural activities. The construction of roads may also provide access for further unscheduled timber extraction and illegal hunting, sometimes accompanied by shifting cultivation and slash and burn practices (Goodland *et al.*, 1991). A major problem with this form of activity is the fact that in tropical forest systems most of the nutrients are stored in the trees as opposed to the soil and small-scale farmers are forced to become nomadic in order to survive. Even if, as is argued below, the influence of logging companies does not represent the primary cause of deforestation, it can create significant conflicts of interest between indigenous groups and external investors.

Colchester (1989: 42) cites an example from Sarawak, where Penan tribesmen petitioned the government to outlaw logging and to return the traditional territories from which they had become displaced, through the following appeal:

> Stop destroying the forest or we will be forced to protect it. We have lived here before any of you outsiders came. We fished in clean rivers and hunted in the jungle ... our life was not easy, but we lived it in content. Now, the logging companies turn rivers into muddy streams and the jungle into devastation. The fish cannot survive in dirty rivers and wild animals will not live in devastated forests.'

Such feelings may be widespread and, equally, they may be largely ignored so as not to compromise sources of foreign revenue. In the case of the Penan the latter was true as their appeals fell on deaf ears. As will be argued in later sections, however, the issue of displacement is not merely caused by exploitation but also by preservation as certain examples from protected areas may attest.

Whilst problems originating from logging generate significant media attention and are portrayed by many to provide the major impetus for deforestation, pressures associated with poverty and international debt (World Bank, 1998) should not be discounted. Indeed, the FAO calculated that 85% of tropical timber was utilized locally as fuelwood (Westoby, 1987). The International Panel on Forests (IPF) stated that the main pressures for deforestation could be attributed to poverty and underdevelopment (Rainforest Foundation International, 1997), particularly in meeting the food requirements of burgeoning populations (Rowe *et al.*, 1992) as forests give way to fields (Barraclough and Ghimire, 1995).

What becomes evident from this debate is that a plethora of demands contributes to the continuation of deforestation practices and that the existence value attributed to their unique ecosystems and stocks of biodiversity is not sufficient to offset fully the opportunity costs associated with the phasing out of other uses. Furthermore, it is difficult to attach an economic value to support the protection of rainforests which in this case are valued for their non-market benefits. This suggests that calls for protection rather than exploitation need to be supported by some form of income, yet in many developing nations this is unlikely to be available. The actual designation of areas as protected has created a major resource for tourism, one which in certain cases may be threatened through over use.

Designation of Protected Area Status

The conveyance of protected area status upon sites of particular environmental and ecological quality represents a conventional conservation option for countering deforestation. Organized and overseen in the international arena by the IUCN (World Conservation Union) a variety of categories of protection are classified (IUCN, 1994). These range from strict conservation (Category I) to managed resource protection areas (Category VI) which seek to balance conservation with development and local community needs. The latter category is similar to the notion of Biosphere Reserves introduced in the 1970s under the UNESCO-sponsored Man and Biosphere programme. Additional protected areas have been designated at a national level in some states, whilst a host of other designations and international conventions lengthen the list further (e.g. World Heritage Sites, Ramsar; Nelson, 1991). Whilst the stricter forms of protected status have traditionally concentrated solely upon preserving the physical resource in an intact and largely unaltered state, many important social and cultural processes have been compromised as a consequence.

> When the whites first arrived in this area, they thought we were wild animals and chased us into the forest. Now that they have found out that we are people they are chasing us out again.
>
> (Okiek hunter-gatherer, Mau Forest, Kenya, 1992, quoted in IIED, 1994)

This is not necessarily an isolated incident but a commonly experienced conflict between conservation interests and local people. Pearce (1997: 12) attests to this when he describes the problems faced by human rights groups representing native people in areas in which environmental interests have 'made pacts with governments with repressive reputations in order to secure protected areas'. In such cases, protection needs to be linked more closely with local needs if conflicts of interest and gross inequities are to be avoided.

Under the aegis of sustainable development and its recognition of such inadequacies, the way in which the environment is conceptualized and the means by which its protection is undertaken has undergone a fundamental change of emphasis in recent years, bringing its multiple facets much more to the fore. Rather than restrict understanding of the environment as being the product of a set of natural processes by way of their effects upon the physical fabric of a given land resource, a more holistic view has prevailed. This new paradigm elevates the cultural and social elements of the environment to be considered simultaneously with the physical resource as primary factors in determining appropriate practices. Under this rationale the fate, and indeed the role, of local people in the design of protected area programmes is viewed as a prerequisite for success. This reflects a growing move towards community-led development and partnership responses to change evident at an international scale (Evans, 1994). In order fully to realize this ideal it has been deemed important that new forms of income be sought to support fragile areas and the activities of indigenous peoples in sustainably utilizing and protecting their local resource base. Tourism has emerged as a key activity in this respect, one which arguably enjoys a symbiotic yet conflicting relationship with many protected areas.

Tourism and Conservation – a Symbiotic Relationship?

A main motivation behind ecotourism development is the employment of tourism receipts to fund protection measures and sustain the resources which represent a magnet to tourists. This notion provides an important basis for ecotourism development, an issue which will be considered in depth below. Additionally, it can provide income by which organizations like Conservation International can finance their innovative 'debt for nature swaps'. In this example, a percentage of a nations foreign debt is purchased in return for conservation agreements from the recipient government. The actual designation of protected area status may in itself generate increased visitor interest (Johst, 1982). This assertion is contested by McCool (1995) but has considerable support in the work of Urry (1995) who introduces the existence of a 'place myth' which can develop around certain protected landscapes. For example, a National Park label can become confused as a certificate of excellence (Evans, 1998), one which conveys a collectable quality to the resource and sets it aside as a place worth visiting. As argued previously in this chapter, the search for authenticity and wilderness, arguably as an adjunct to growing environmental consciousness, combined with an enhanced opportunity to travel, has brought previously inaccessible environments within the scope of the modern-day tourist.

Increasing emphasis upon sustainable tourism and the attraction of nature has culminated in the advent of ecotourism and its acknowledgement as one of the fastest growing sectors of the industry. This has undoubtedly been picked up by the tourism companies who are constantly seeking new marketing handles by which to promote such forms of specialist travel. Unsurprisingly, the issue of protected area status has not escaped the attention of the marketeers:

> Designation as a UNESCO World Heritage Site confers upon a location the highest accolade available in terms of importance and significance to our planet, the loss of which would constitute a serious decline in the Earth's cultural and natural wealth. With the exception of Papua New Guinea, each and every one of our destinations possesses at least one World Heritage Site, most of which can be visited in one or more of our itineraries.
>
> (Reef and Rainforest Tours, 1999)

What was initially considered as a lifeline for threatened resources has, through the passage of time, created its own forms of deterioration, suggesting that the efficacy of using tourism as a form of environmental management is very much open to question. This is undoubtedly a debate which warrants deeper investigation and discussion.

Ecotourism and Tropical Rainforests

What then is ecotourism and how does it relate to the environmental management of tropical forests? The notion of ecotourism has developed out of a growing interest in travelling to more natural environments, itself a manifestation of the renewed interest internationally in environmental and nature concerns (Ceballos-Lascurain, 1996). Ecotourism has been defined as the act of 'travelling to relatively undisturbed or uncontaminated natural areas with the specific objective of studying, admiring and enjoying the scenery and its wild plants and animals, as well as any existing cultural manifestations (both past and present) found in these areas' (Boo, 1990: xiv). A further, and perhaps more focused definition is provided by Fennell (1999: 43) who describes ecotourism as a 'sustainable form of natural resource-based tourism that focuses primarily on experiencing and learning about nature, and which is ethically managed to be low-impact, non consumptive, and locally oriented (control, benefits and scale). It typically occurs in natural areas, and should contribute to the conservation or preservation of such areas.' In each case, the tropical rainforests provide an obvious venue and a means by which to contribute positively to their protection.

This form of development has proved successful in many tropical environments, a commonly cited example of which has been the role of tourism in conserving the silver-backed gorillas of Rwanda. Heralded as a success story by the Rainforest Action Network (1998), evidence suggests that the Mountain Gorilla Project (MGP) set up in the late 1970s had derived tourism receipts of US$1 million per annum in 1989, which had provided a profit on running the park (Lindberg and Huber, 1993). Before this project was established, the mountain gorilla was fast becoming a threatened species, hunted by poachers and slaughtered as pests by local farmers (Weber, 1993). There has been a flip side to this story, however, not least in the way that certain animals become used to humans and trusting of their actions. Fears that a drying up of tourism revenues as a result of political disruptions in Rwanda and its surrounding territories will terminate the MGP are accompanied by the realization that they may again become endangered species once hunting resumes. Slightly less serious although problematic is the actual access of visitors to such fragile ecosystems and the potential conflicts which may arise from chance meetings with illegal hunters. Ham (1995: 3) describes 'trackers carry[ing] machetes to cut a narrow path through the forest ... [of a] ranger carrying an AK47 assault rifle in case of an encounter with armed poachers'.

The impacts of visitors trekking through virgin forests armed with machetes raises a number of questions, not least the extent to which people are really conforming to the 'take only photographs, leave only footprints' motto adopted by the ecotourism lobby (Rowe *et al.*, 1997) in areas characterized by their fragile ecosystems. Much has been written about ecotourism in recent years (e.g. Mowforth and Munt, 1998; Fennell, 1999) and a number of common themes have emerged from the literature. Firstly, the optimum number of ecotourists is less than that for conventional forms of mass tourism and thus the unit cost per visitor is greatly increased. Mowforth and Munt (1998: 125–126) point out that this has led to an emergence of a new tourist class, one which corresponds closely to the commonly perceived model of the environmentally conscious citizen. The fear that as previously remote, undisturbed regions become 'discovered', these tourists will continue to seek authenticity is difficult to ignore.

This will not only have implications for physical environments but also for cultural and social processes, an issue which is already drawing attention, not least in the theme of staged authenticity or zooification (Mowforth and Munt, 1998: 273). Colchester (1994: 3) describes the existence of ' "enforced primitivism" whereby indigenous people are accommodated in protected areas so long as they conform to the stereotype and do not adopt modern practices'. Returning to the Penan tribe, described earlier in the chapter for their anti-logging appeals, and this situation becomes of even more concern. Within the Mulu National Park in Sarawak, visitors are provided with an opportunity to observe the Penan way of life, with the construction of a traditional Malaysian longhouse representing the centrepiece of the attraction. In terms of authenticity, however, this is unfortunate as the Penan tribe have never lived in such buildings which are the traditional dwellings of the Iban people (Survival International, 1991).

In an attempt to attract the ecotourist, a range of activities and attractions are constantly being sought. In certain cases the rainforest itself may not be enough to stimulate visitor interest throughout the duration of a vacation period, and the notion of adding value to the resource becomes important. Additionally, although for some driven by romantic images drawn from Tarzan movies, the discerning ecotourist is evolving to demand certain standards of sanitation and security. In the words of one visitor to the rainforests of central America, 'I want to know the creepy crawlies are here but I don't want to live with them'. This attitude has undoubtedly influenced the form in which ecotourism destinations have been planned and managed, with the introduction of basic facilities and guided treks giving way to boardwalks and *en route* chalets and toilet blocks. This does not necessarily incur the wrath of the environmentalists who may indeed support such developments as being preferable to unregulated access to wilderness areas armed with machetes to hack down any vegetation blocking their way. The following illustrative case study is drawn from a recent research visit the author undertook in March 1999 to the Sultanate of Brunei Darussalam.

Ulu Temburong National Park, Brunei Darussalam

Located on the north-west tip of Borneo, bordered by the east Malaysian states of Sarawak and Sabah, the Sultunate of Brunei Darussalam covers an area of 5765 square kilometers containing a population of only 275,000 people. A nation which has avoided problems associated with foreign debt due to its abundant reserves of oil, it has managed to conserve its rainforests, a resource which covers an estimated 80% of the nation. Around 60% of this resource is classified as being unaffected by human activity (Government of Brunei, 1996). Despite avoiding the worst of the South-east Asian economic crises, Brunei has nevertheless suffered recession in recent years and, in an effort to diversify its economic base, the issue of tourism has been identified as a major new sector of development (Government of Brunei, 1997). Unsurprisingly, considering the structure of the country, the theme of ecotourism has been identified as a primary focus of this process.

Situated to the north-east of the country, Ulu Temburong National Park is effectively cut off by road from much of Brunei by Sarawak. Only accessible by river, access to the site is only possible by longboat. Some 90% of the land here consists of primary tropical high forest with Iban communities still living in their traditional longhouses. Regular trips are organized to the National Park through Sunshine Borneo Tours who enter agreements with the Forest Service to accommodate tourist groups. The Brunei Government, with sponsorship from Brunei Shell, have developed a Rainforest Field Studies Centre in Kuala Belalong, within the park's boundaries, to provide research and educational opportunities for both university and secondary school levels. Additional facilities are provided for tourist groups through the Forest Service. These include refreshments and toilet facilities.

Brunei is a nation which remains in its infancy in terms of tourism development, with existing markets dominated by business visitors. In an attempt to broaden this base and indeed to attract some of the business tourists to travel more widely in Brunei, a tourism infrastructure has been developed in Ulu Temburong. This consists of a 7 km continuous section of constructed boardwalks, bridges and stairways. Interpretation boards are located along the trail, identifying the different tree and plant species present in the region. At regular intervals, rest and toilet facilities are provided. Tourists groups can travel at their own pace and are accompanied by an experienced guide provided by the Forest Service.

The centre piece of the attraction is a 62 m high metal structure which enables visitors to negotiate a series of towers and walkways above the level of the canopy. This affords unparalleled views of the forest and is promoted as an ultimate photographic opportunity. The guide explained to the visitors the importance of the raised structures in reducing negative impacts within the forest itself. Additionally, he pointed out, the existence of such permanent structures enabled the National Park to accommodate greater numbers of visitors. Due to the economic stability of the country, the need to redress deforestation pressures has not been an issue and tourism receipts have never been considered important sources of funding for conservation activities. In terms of an authentic rainforest experience, however, the artificial structures lessened the extent to which the visitors considered the area as wilderness.

This form of development is by no means restricted to Ulu Temburong, however, and the commodification of ecotourism internationally into a series of universally designed routeways and cabins may already be affecting the distinctiveness of different nations and their resources. A recent recipient of a British Airways sponsored 'Tourism for Tomorrow' Award was a rainforest enterprise in Ghana, one which used similar structures to those encountered in Brunei. Costa Rica, itself a prime ecotourism site, has perhaps gone one step further with the introduction of a canopy ride in a cable car. In each case, increasing numbers of ecotourists are attracted to rainforest resources, with the additional income generated through their visitation enabling further 'improvements' to the attraction. The example from Brunei is perhaps unique in the way that it has arguably been able to bypass the initial stages of ecotourism development due to its absence of foreign debt and externally influenced pressures for deforestation. The case of Costa Rica has been somewhat different, yet arguably it has produced a similar result.

The Case of Costa Rica

Application of the same debates to Costa Rica provides an interesting contrast to the Brunei example. Whilst Brunei has utilized its rainforest resource as a primary attraction in an attempt to stimulate tourism activity nationwide, Costa Rica has developed somewhat differently. Following significant levels of 20th century deforestation, the rainforests of Costa Rica represent only a fragment of the originally extensive Central American resource. Declining to just 11% total land cover, all of which falls under some category of protection, ecotourism has rapidly emerged as the nation's primary development sector. From a base of 50,000 visitors to the Park system in 1986, tourism activity increased dramatically to reach a level of 250,000 people by 1991 (Weaver, 1997). In this case, Costa Rica has been described as both a success story and as a victim of that success (Rainforest Action Network, 1998), with the subsequent introduction of controls and visitor ceilings required in order to avoid resource deterioration. Whilst Brunei has sought to stimulate its tourism market using the rainforest as a primary attractor, Costa Rican forests have often been seen as secondary resources. A nation renowned for its coastal resources and Mayan heritage, a significant tourism infrastructure has developed around its beach resorts. This has provided a market within which to promote its rainforests as an additional, though not necessarily central, attraction. The emergence of a considerable day excursion visitor base has placed increasing pressure upon the natural resources of the country (Wood, 1993).

Whilst the genesis of ecotourism in the two examples differs, the resulting architectural and infrastructural development shows considerable uniformity. The boardwalks and canopy towers, alongside the toilet and rest facilities have, in each case provided a safe, sanitized environment for people to visit. This issue is of utmost importance when considering the future demands which will affect rainforest resources globally.

Conclusions

This chapter began by setting out a situation in many developed nations whereby the regulation of access to forest resources was based increasingly upon built or artificial structures, to some extent compromising the naturalness of the visitor experience. It was implied that this system was introduced as a form of environmental management by which many of the negative impacts associated with access could be minimized. This also reflected a climate of increased environmental awareness across society as a whole and concomitant demands to visit areas of natural attractiveness and biodiversity.

This argument was then countered by a different situation arising in many tropical regions, whereby tourism was proposed as an environmental management technique in its own right. Suffering significant pressures for deforestation, the notion of tourism receipts as a funding mechanism for ensuring conservation and protection of natural resources prevailed under the theme of ecotourism. Utilizing a vision of small, committed groups of visitors, driven by altruistic motivations to play their role in managing valuable resources, ecotourism was considered by many as a potential panacea to traditional forms of resource exploitation. Linked to the issue of protected area designation, the above stated views have not necessarily been borne out in reality, with destinations taking on a collectable quality and, once adequately patronized, being subject to increasing forms of regulation and control. This process may ultimately lead to a fragmentation of the existing ecotourism market, with the security and familiarity provided by artificial structures, the emerging architecture of rainforest tourism, attracting a new form of mass tourist, with the more adventurous visitors constantly seeking new, unspoiled destinations in ever more remote regions of the world.

The issue of ecotourism as a form of environmental management in its own right deserves further investigation. Whilst at the start of tourism development this may to some extent be the case, it would appear, however, that it may in fact represent an initial phase of a longer process more akin to developed nations, rather than an end in itself. In the long term, it is difficult to imagine the theme of ecotourism continuing in its present guise. It may become viewed increasingly as a dynamic concept, one which needs to be allied more closely to the full range of economic, social, cultural and physical factors which dictate the broader nature and management of natural resources internationally. Whilst calls to promote and protect diversity represent a key motivation behind the notion of sustainability, a situation by which local distinctiveness is compromised by the need to accommodate growing numbers of visitors may prove be a consequence of current practices.

Returning to the analogy provided by forest resource management in much of the developed world, it may be the case that areas containing the infrastructure required to accommodate larger scale tourism access may ultimately be sacrificed as 'honeypots' whilst other, more fragile sites, develop as small-scale, low-impact and increasingly local community-controlled enterprises. Perhaps the most important task of environmental management, and thus tourism development, into the future will be its ability to control the pace of change. The scenario adopted by this work thus places the notion of ecotourism within a temporal stage of dynamic tourism development, suggesting that the time scale may prove to be more important than geographical setting in dictating the future form of environmental management both practised and required. Safeguarding cultural as well as biological diversity is essential, making certain that indigenous traditions are not compromised as a selling point for the destination.

References

Arnold, J. (1992) Community forestry: ten years in review. Community Forestry Note No. 7, Forest Trees and People Programme, FAO, Rome.

Barraclough, S. and Ghimire, K. (1995) *Forests and Livelihoods*. St Martins Press, New York.

Benson, J. and Willis, K. (1991) The demand for forests for recreation, Paper No. 6. In: *Forestry Expansion: A Study of Technical, Economic and Ecological Factors.* Forestry Commission, Edinburgh.

Boo, E. (1990) *Ecotourism: The Potentials and Pitfalls.* World Wildlife Fund (WWF), Washington D.C.

Brown, S., Hall, C., Knabe, W., Raich, J., Trexler, M. and Woomer, P. (1993) Tropical forests: their past, present and potential future role in the terrestrial carbon budget. *Water, Soil and Air Pollution* 70, 71–94.

Ceballos-Lascurain, H. (1996) Tourism, ecotourism and protected areas. *IV World*

Congress on National Parks and Protected Areas, IUCN (World Conservation Union), Gland, Switzerland.

Ciesla, W. (1995) Climate change, forests and forest management: an overview. FAO Forestry Paper 126, United Nations, Rome.

Colchester, M. (1989) *Pirates, Squatters and Poachers: The Political Ecology of Dispossession of the Native Peoples of Sarawak*. Survival International, London.

Colchester, M. (1994) Salvaging nature: indigenous peoples, protected areas and biodiversity conservation. Discussion Paper, UNRISD, Geneva.

Cooper, C., Fletcher, J., Gilbert, D. and Wanhill, S. (1998) *Tourism Principles and Practice*, 2nd edn. Addison Wesley Longman, Harlow.

Evans, S. (1992) Tourism motivations in forested environments. Unpublished Working Paper, Anglia Polytechnic University, Chelmsford.

Evans, S. (1994) The community forest initiative: past, present and future perspectives. In: Fodor, I. and Walker, G. (eds) *Environmental Policy and Practice in Eastern and Western Europe*. Centre for Regional Studies, Hungary, pp. 207–214.

Evans, S. (1998) Community forestry: countering excess visitor demands in England's national parks. Paper delivered to 9th International Symposium on Society and Resource Management, University of Missouri, Columbia, May.

Fennell, D. (1999) *Ecotourism: An Introduction*. Routledge, London.

Gomez-Pompa, A., Vazquez-Yanes, C. and Guevara, S. (1972) The tropical rainforest: a non-renewable resource. *Science* 117, 762–765.

Goodland, R., Asibey, E., Post, J. and Dyson, M. (1991) Tropical moist forest management: the urgency of transition to sustainability. In: Costanza, R. (ed.) *Ecological Economics: The Science and Management of Sustainability*. Columbia University Press, New York, pp. 486–515.

Government of Brunei (1996) *The Treasures of Brunei Darussalam*. Health, Safety and Environment Department, Bandar Seri Begawan.

Government of Brunei (1997) *Explore Brunei*. Ministry of Tourism, Bandar Seri Begawan.

Ham, M. (1995) *Cashing in on the Silver Backed Gorilla*. Ecotourism Special, Panos Institute, London.

IIED (1994) Whose Eden: an overview of community approaches to wildlife management. A Report to the Overseas Development Administration of the British Government, International Institute for Environment and Development, London.

Irving, J. (1987) *The Public In Your Woods*. Land Decade Education Council, London.

IUCN (1994) *United Nations List of National Parks and Protected Areas*. World Conservation Union, Gland, Switzerland.

Johst, D. (1982) *Does Wilderness Designation Increase Recreation Use?* Bureau of Land Management, Washington D.C.

Lindberg, K. and Huber, R. (1993) Economic issues in ecotourism management. In: Lindberg, K. and Hawkins, D. (eds) *Ecotourism: A Guide For Planners and Managers*. The Ecotourism Society, Vermont.

McCool, S. (1995) Does wilderness designation lead to increased recreational use? *Journal of Forestry*, January, 39–41.

Madeley, J. and Warnock, K. (1995) Biodiversity: a matter of extinction – the challenge of protecting the South's biological heritage. Media Briefing No. 17, Panos Institute, London.

Mowforth, M. and Munt, I. (1998) *Tourism and Sustainability: New Tourism in the Third World*. Routledge, London.

Myers, N. (1989) *Deforestation Rates in Tropical Forests and their Climatic Implications*. Friends of the Earth, London.

Nelson, J. (1991) Sustainable development, conservation strategies and heritage. In: Mitchell, B. (ed.) *Resource Management and Development*. Oxford University Press, Oxford.

Pearce, F. (1997) People and parks: wildlife, conservation and communities. Media Briefing No. 25, Panos Institute, London.

Pearce, F. (1998) Beyond hope. *New Scientist*, 31 October, 17–18.

Peters, C., Gentry, A. and Mendelsohn, R. (1989) Valuation of an Amazonian rainforest. *Nature* 339, 655–656.

Phillips, O., Hall, P., Gentry, A., Sawyer, S. and Vazquez, R. (1994) Dynamics and species richness of tropical rainforests. In: *Proceedings of the National Academy of Sciences, USA*, 2805–2809.

Rainforest Action Network (1997) *Facts about Rainforests, Factsheet 1D*. http:www.ran.org/ran/info.

Rainforest Action Network (1998) *Can Ecotourism Save the World's Forests*. RAN, Lismore.

Rainforest Action Network (1999) *Amazon Oil Campaign, Factsheet*. http://www.ran/info_center/factsheets.

Rainforest Foundation International (1997) Report on the International Panel on Forests, New York.

Redclift, M. (1987) *Sustainable Development: Exploring the Contradictions*. Methuen, London.

Reef and Rainforest Tours (1999) *Spring Newsletter*, Sussex.

Repetto, R. (1990) Deforestation in the tropics. *Scientific American* 262, 36–45.

Rowe, D., Leader-Williams, M. and Dalal-Clayton, B. (1997) *Take Only Photographs, Leave Only Footprints: The Environmental Impacts of Wildlife Tourism*. International Institute for Environment and Development (IIED), London.

Rowe, R., Sharma, N. and Browder, J. (1992) Deforestation: problems, causes and concerns. In: Sharma, N. (ed.) *Managing the World's Forests*. Kendall Hunt, Iowa, pp. 33–45.

Skole, D. and Tucker, C. (1993) Tropical deforestation and habitat fragmentation in the Amazon: satellite data from 1978 to 1988. *Science* 260, 1905–1909.

Survival International (1991) *Tourism: Special Issue, Survival No. 28*. Survival International, London.

UNCED (1992) Resolution 44/228, United Nations Conference on Environment and Development, New York.

Urry, J. (1995) *Consuming Places*. Routledge, London.

Vivian, J. (1992) Foundations for sustainable development: participation, empowerment and local resource management. In: Ghai, D. and Vivian, J. (eds) *Grassroots Environmental Action: People's Participation in Sustainable Development*. Routledge, London, pp. 50–80.

WCED (1987) *Our Common Future*. World Commission on Environment and Development. United Nations, New York.

Weaver, D. (1997) Ecotourism in Costa Rica. In: France, L. (ed.) *The Earthscan Reader in Sustainable Tourism*. Earthscan, London, pp. 205–207.

Weber, W. (1993) Primate conservation and ecotourism in Africa. In: Potter, C., Cohen, J. and Jankzewski, D. (eds) *Perspectives on Biodiversity: Case Studies of*

Genetic Resources. American Association for the Advancement of Science, Washington, DC, pp. 129–150.

Westoby, J. (1987) *The Purpose of Forests: Follies of Development*. Blackwells, Oxford.

Winterbottom, R. (1990) *Taking Stock: The Tropical Forest Action Plan After Five Years*. World Resources Institute, Washington, DC.

Wood, M. (1993) Costa Rican parks threatened by tourism boom: society launches letter-writing campaign. *The Ecotourism Society Newsletter* 3, 1–2.

World Bank (1998) World Development Report 1998–1999, Table 21: Aid and Financial Flows, http://www.worldbank.org/wdr/wdi/wdi21.pdf.

World Resources Institute (1997) *The Last Frontier Forests: Ecosystems and Economies on the Edge*. WRI, Washington, DC.

9

Wilderness Management in the Forests of New Zealand: Historical Development and Contemporary Issues in Environmental Management

C. Michael Hall and James Higham

Introduction

New Zealand's 'clean and green' image is an essential element of its attractiveness as an international tourism destination. Although New Zealand's rural environment, based on images of sheep grazing on rolling green hills, is an important contributor to this image, it is primarily based on perceptions of unpolluted rivers and lakes, alpine areas and forest wilderness which centre on an extensive system of conservation lands (Higham, 1996; Hall *et al.*, 1997).

Protected areas in New Zealand include national, maritime and forest parks and three World Heritage Areas (Tongariro, Southwest New Zealand Te Wahipounamu and the Sub-Antarctic Islands World Heritage Areas). These designations are known generically as the 'Conservation Estate'. Visitor interest in the Conservation Estate offers both an important source of foreign exchange earnings and the challenge of sustainable environmental management as tourists are increasingly attracted to locations that are highly valued by virtue of their outstanding natural qualities. These are areas of high conservation value and are susceptible to impact.

A series of ten Wilderness Areas lie at the core of New Zealand's National and Forest Parks system. New Zealand's Wilderness Areas are extensively forested (Table 9.1). The two exceptions to this rule are the Olivine and Garvie Wilderness Areas. These areas are predominantly characterized by ice plateau and mid- to high-alpine tussock respectively. Wilderness Areas in New Zealand comply with a highly purist legislated definition of Wilderness. They are set within a wider extensively forested Conservation Estate which provides qualities of wilderness experience for the majority of domestic and international recreational visitors (Kearsley, 1983; Higham, 1996).

The forest wilderness has long played a part in New Zealand's economic development. For much of the 19th and early 20th century it was the source of wood for housing, mining and industrial purposes. As attitudes towards the wilderness have changed so the majority of native forest areas came to be protected from timber cutting. However, its economic use value still determines the means by which the forest wilderness is managed (Wynn, 1977).

The importance of the forest wilderness to contemporary economic development in New Zealand was confirmed by the publication of *New Zealand Conservation Estate and International Visitors* in 1993 by the Department of Conservation (DOC), the government authority responsible for the Conservation Estate, and the New Zealand Tourism Board (NZTB), which is responsible for the promotion of New Zealand to international tourists (DOC/NZTB, 1993). The document established the policy framework within which the Conservation Estate is managed for tourism and conservation purposes. This policy setting aims at maximizing the economic benefits

Table 9.1. The forest character of New Zealand's ten proposed Wilderness Areas.

Wilderness designation	Area (ha)	Forest character
Raukumara	44,000	Densely forested Raukumara Range, East Cape, North Island. Motu River flows through heavily forested gorges for 100 km.
Kaimanawa	47,000	Comprises native beech forest, sub-alpine tops, tussock basins and varied riverscapes.
Tasman	94,000	Forests of silver and mountain beech dominant throughout. Sub-alpine tops covered by red tussock and scrub species. Dense stands of nikau palms along the coast between the Heaphy and Kohaihai rivers.
Paparoa	36,000	Four beech forest species predominate within this wilderness area. Dense sub-alpine scrub.
Adams	54,000	Gorges, icefalls, glaciers and dense sub-alpine scrub dominate this wilderness. The Poerua State Forest comprises 25% of the Adams Wilderness Area.
Hooker	44,000	Large rivers and grass river flats separated by densely forested bluffs, gorges and bush covered necks.
Olivine	55,000	Comprises the Olivine Ice Plateau. Silver beech generally forms the bushline with red beech dominant in higher areas of this Wilderness Area. Some undulating tussock plateau.
Garvie	43,000	Mid- to high-alpine tussock grassland.
Waitutu	30,000	Relatively open podocarp/beech forest on gentle topography.
Pegasus	63,000	Almost entirely forested island Wilderness Area. Varied vegetation that exists right to the rugged coastline.

Source: Federated Mountain Club (1981).

of tourism, a philosophy that in this chapter is referred to as 'New Economic Conservation'. The aims and trends emerging from this policy document included the doubling or tripling of international tourist arrivals to New Zealand and increases in demand for forest and national park recreation resources commensurate with this goal. These trends were described by Boas (1993) as 'disturbing' given the absence of a framework for environmental management.

Pressures of recreational use (and associated demands for the development of access and other tourist infrastructure) on the Conservation Estate have and continue to intensify. This arises from both increasing domestic use of the Conservation Estate, coupled with fluctuating, but generally rapid growth in inbound tourism in the last decade. Arising from the growth of overseas interest in New Zealand's wilderness and forested areas has been evidence of ecological impact (Kearsley and Higham, 1997), the degradation of wilderness values (Kearsley, 1997) and increasing perceptions of crowding (O'Neill, 1994; Higham, 1996). Reduced visitor satisfaction by overseas visitors is a logical implication if this is the case, but so too is the likelihood of degraded forest resources and lowering of wilderness values. The management of forest wilderness resources has therefore become a pressing concern in light of current and projected growth in visitor markets.

Through this sequence, the demand-side of wilderness and forest-based recreation and tourism in New Zealand has been the focus of various researchers (see Chapter 5). It may be argued, however, that the sustainable management of tourism in forest wilderness environments necessitates that consideration also be given to the supply-side of wilderness management in both the historical and contemporary contexts. This chapter therefore proposes that the current management of New Zealand's forest wilderness environments, in relation to tourism, needs to be understood in the context of institutional arrangements that have developed over time, thereby providing a changing framework for the management of tourism in the forest wilderness.

The Study of Institutional Arrangements

Institutional arrangements provide a set of rules and procedures that regulates how and where demands on public policy can be made, who has the authority to take certain decisions and actions, and how decisions and policies are implemented. Institutions are 'an established law, custom, usage, practice, organisation, or other element in the political or social life of a people; a regulative principle or convention subservient to the needs of an organized community or the general needs of civilization' (Scrutton, 1982: 225). The study of institutional arrangements has long been regarded as a significant aspect of resource and environment management (e.g. Mitchell, 1989). For example, O'Riordan (1971: 135) observed that:

> One of the least touched upon, but possibly one of the most fundamental, research needs in resource management is the analysis of how institutional arrangements are formed, and how they evolve in response to changing needs and the existence of internal and external stress. There is growing evidence to suggest that the form, structure and operational guidelines by which resource management institutions are formed and evolve clearly affect the implementation of resource policy, both as to the range of choice adopted and the decision attitudes of the personnel involved.

More recently, the study of institutional arrangements has been seen as critical to understanding the conditions by which sustainable development may be encouraged (e.g. Ostrom, 1986). However, while the study of institutional arrangements is fundamental to the context within which environmental and resource management occurs, little consideration has been given to the role that institutional arrangements play with respect to tourism and environmental management (Hall and Jenkins, 1995). There is therefore relatively little understanding of the way in which the institutional context has influenced the environmental management of tourism.

The following discussion seeks to address the significance of institutional arrangements for the environmental management of tourism in the context of the New Zealand forest wilderness. The chapter adopts an historical perspective and outlines the way in which previous sets of institutional arrangements and values have established environmental management regimes which influence subsequent policy settings and actions. As the subsequent sections argue, four main periods of institutional arrangements can be identified with respect to forest wilderness management in New Zealand. The chapter concludes by noting the implications of these arrangements for contemporary environmental management practice and the future of the forest wilderness.

Utilitarian conservation (1870s–1940s)

The history of New Zealand's protected area system highlights the role of a utilitarian conservation ethic in the development of the national and forest parks. The nucleus of present-day Tongariro National Park, New Zealand's first National Park, was gifted to the New Zealand Government in 1887, with the park finally legally designated in 1894. The considerable delay between the deeding of the land of the Maori Chief Te Heuheu Tukino to the Crown, and the actual establishment of the park, reflected the Government's concern that only 'worthless' land would be incorporated in the park. 'There had to be absolute certainty that land being added to the park had no economic value' (Harris, 1974). In speaking to Parliament on the proposed park, the Hon. John Balance (New Zealand, 1887: 399) stated: 'I may say that this land is particularly suited for a national park. It has all the appearance of a park in itself, and many persons, looking at it, would imagine it had been laid out artificially, and created at enormous expense for the purpose of a park.'

The 'worthless lands' view of national parks, so characteristic of early attitudes towards parks in Australia, Canada and the United States (Runte, 1972, 1973, 1979; Hall, 1988a, 1998), was also dominant in New Zealand (Hall, 1988b; Hall and Shultis, 1991). For example, in discussing Tongariro National Park, the Hon. John McKenzie, Minister for Lands, was reported as telling Parliament that, 'anyone who had seen the portion of the country ... which he might say was almost useless so far as grazing was concerned would admit that it should be set apart as a national park for New Zealand' (New Zealand, 1894: 579). In a similar fashion to Canada and the United States, the New Zealand Government saw national parks as a means to develop areas economically through tourism, the aesthetic values of regions being the attraction to the tourist. To quote Ballance again on the subject of Tongariro National Park: 'I think that this will be a great gift to the colony: I believe it will be a source of attraction to tourists from all parts of the world and that in time this will be one of the most famous parks in existence' (New Zealand, 1887: 399).

As in North America, lodges were established within national parks to provide comfortable surroundings for the well-heeled visitor. Similarly, railways played a prominent role in bringing visitors to the parks (Hall, 1988a). A degree of protection for forest wilderness areas was only a fortunate by-product of national park declaration not a primary cause, with activities of relatively small numbers of visitors only having highly localized impacts on the environment. However, a degree of protection for indigenous flora and fauna did arise from the activities of scientific societies and concerned natural historians (Thomas, 1891), including the reservation of several areas as bird and fauna sanctuaries (Wynn, 1977). Nevertheless, the dominant attitude towards national parks in New Zealand was that they were provided for the purposes of recreation and tourism, and not for the preservation of untrammeled nature. The precarious existence of nature in New Zealand was enough to provoke John Muir into commenting in 1904 that the New Zealand Government was 'selling its country's wealth', its forests, 'for a leg of mutton' (unpublished Muir diary entry in Hall, 1993).

The introduction of alien species into the parks of New Zealand is a recurring theme in New Zealand's park history. Sheep were not the only introduced species to create problems of land degradation. The introduction of deer for hunting purposes caused enormous damage to forested areas. The New Zealand public's opposition to the introduction of new species was exemplified by the widespread indignation and opposition to a Lady Liverpool's efforts to introduce grouse into Tongariro Park. Professor H.B. Kirk, one of New Zealand's leading natural historians, sent an angry letter to the *Evening Post*, which had earlier applauded Lady Liverpool's efforts as likely to 'give added attractions to sportsmen coming to New Zealand from the Old Country'; 'No other country would do so ludicrous a thing as to convert the most distinctive of its national parks into a game preserve ... this thing is an insult to the Maori donors and to all lovers of New Zealand as New Zealand.' Kirk's letter appeared to find a supportive response in a wide range of individuals and authorities. By the end of 1924 the New Zealand Legislative Council had 'pushed through a resolution condemning all introduction and proclaiming that the park should be held inviolate' (in Harris, 1974: 109–110). Nevertheless, despite the emergence of a positive ecological viewpoint towards the role of the National Parks it should be noted that they were still seen primarily in economic terms.

The creation of a New Zealand National Parks system (1950s–1960s)

Until 1952 the National Parks of New Zealand were established under a variety of Acts and each park was managed separately. Parks were created because of their spectacular scenery. The parks contained examples of New Zealand's unique fauna and flora, but also introduced plant and animal species, such as deer, pigs, goats and opossums, which caused enormous damage to the forest environment (Veblen and Stewart, 1982). The lack of a coordinated and systematic approach to national park planning was criticized by a number of groups and individuals in the inter-war years including the Director of Kew Gardens, Dr Arthur Hill (Thompson, 1976; Fleming, 1979). These criticisms had little impact on government policy towards forest conservation, however, the recreational perspectives of tramping associations and mountaineering clubs did have a significant impact on park policy. Through a comparison of overseas initiatives in park planning a Federated Mountain Clubs (FMC) sub-committee decided 'to put forward suggestions for more systematic general control, based upon the successful and businesslike examples of the United States and Canada' (Thompson, 1976: 9). The lobbying of the FMC appeared to have some influence on the New Zealand Government, but the reorganization of the parks had to wait until after the Second World War.

Ron Cooper, Chief Land Administration Officer of the Department of Lands and Surveys (DLS), played a prominent role in the creation of the National Parks Act 1952 which provided for greater access for visitors to forest wilderness areas. Cooper conceived New Zealand's national parks as wilderness to which the general public should have access.

> A national park is ... a wilderness area set apart for preservation in as near as possible its natural state, but made available for and accessible to the general public, who are allowed and encouraged to visit the reserve. In such an area the recreation and enjoyment of the public is a main purpose, but at the same time the natural scenery, flora and fauna are interfered with as little as possible. Such a reserve should contain scenery of distinctive quality, or some natural features so extraordinary or unique as to be of national interest and importance, and as a rule it should be extensive in area.
>
> (Ron Cooper, 21 January 1944, in Thompson, 1976: 11)

In line with American perspectives, Cooper had an anthropocentric perception of wilderness in which he saw New Zealand's wilderness areas as being recreational in character and did not see them as scientific reserves. The recreational importance attached to New Zealand's national parks was demonstrated in the National Parks Act 1952 which, following on from North American national park legislation, defined the purpose of the parks as 'preserving in perpetuity ... for the benefit and enjoyment of the public, areas of New Zealand that contain scenery of such distinctive quality or natural features so beautiful or unique that their preservation is in the national interest'.

The extent of the contemporary influence of the North American national park systems on New Zealand is further evidenced by the study tour of these countries by P.H.C. Lucas, Director of National Parks and Reserves, in 1969 (Lucas, 1970). The report of the study tour contains a wide account of park management practices and has many sections entitled 'lessons for New Zealand' (Lucas, 1970). The report served as one of the major determinants in the direction of New Zealand's national park policies through the 1970s and the early 1980s. Similarly, the review of the administrative structure of national parks and reserves in 1979 also showed a great many American influences (Government Caucus Committee Report, 1979).

Wilderness preservation (1970s–1980s)

Legal recognition of the wilderness concept in New Zealand first occurred under the National Parks Act 1952, with the first wilderness area being established under the Act in 1955. The National Parks Act 1980 has similar provisions which repeat the 1952 Act. Provision for wilderness areas was also made in the Reserves and Domains Act 1955 which was subsequently revised as section 47 of the Reserves Act 1977. However, no wilderness area was actually established in the reserves system. Section 14 of the National Parks Act 1980 referred to the creation and management of wilderness areas:

> (1) The Minister may, on the recommendation of the Authority made in accordance with the management plan, by notice in the Gazette, set apart any area of a park as a wilderness area, and may in like manner revoke any setting apart.
>
> (2) While any area is set apart as a wilderness area,
>
> (a) It shall be kept and maintained in a state of nature:
>
> (b) No buildings of any description, ski-lifts, or other apparatus shall be erected or constructed in the area:
>
> (c) No animals or vehicles of any description shall be allowed to be taken into or used or kept in the area:
>
> (d) No roads, tracks, or trails shall be constructed in the area, except such tracks for the use of persons entering the area on foot as are contemplated by the management plan
>
> (National Parks Act 1980)

The New Zealand National Park Authority's (1978: 1.1) policy, as established by the Act, was 'first to preserve the parks and then, so far as the principle of preservation allows, to permit the fullest proper use and enjoyment by the public'. This philosophy was indicated in the four standard park land classifications adopted by the Authority: special area, wilderness area, natural environment area, and facilities area. Under this classification scheme a wilderness area was defined as:

> an area whose predominant character is the result of the interplay of purely natural processes, large enough and as situated as to be unaffected, except in minor ways, by what takes in the non-wilderness around it. In order that the enjoyment of a completely natural unspoilt environment may be experienced, access to and within a wilderness area will be by foot only.
>
> (New Zealand National Parks Authority General Policy, 1978: 3.2)

In 1981, following the 50th Jubilee Wilderness Conference organized by the FMC, the Minister of Lands and Forests established a Wilderness Advisory Group (WAG) to advise the Minister 'on policy for wilderness establishment and use, on the identification and assessment of potential wilderness and on priorities for action' (DLS, 1984: C.1). The FMC proposed ten wilderness areas at the Wilderness Conference which were examined by WAG established later that year (Table 9.1; Hall 1988b). WAG consisted of representatives from both government and public interest organizations. According to WAG (1985: n.p.), 'wilderness areas are wild lands designated for their protection and managed to perpetuate their natural condition and which appear to have been affected only by the forces of nature, with any imprint of human interference substantially unnoticeable'.

> Wilderness areas in New Zealand should meet the following criteria:
> (i) they will be large enough to take at least 2 days' foot travel to traverse;
> (ii) they should have clearly defined topographic boundaries and be adequately buffered so as to be unaffected, except in minor ways, by human influences;
> (iii) they will not have developments such as huts, tracks, bridges, signs, nor mechanical access.
>
> (WAG, 1985: n.p)

WAG (1985) did recognize that 'a wilderness system should have a wide geographic distribution, and contain diversity in landscape and recreational opportunity'.

Provision for wilderness areas on Forest Service land was made by a 1976 amendment to the Forest Act 1949. The first State Forest Park wilderness area was established at Raukumara in June, 1986. As of the beginning of 1987 two further wilderness areas had received ministerial approval and approval-in-principle, respectively. They were the Tasman Wilderness area (91,000 ha) in the North West Nelson State Forest Park (now Kahurangi National Park), and the Paparoa Wilderness (35,000 ha) in the Paparoa Range (now Paparoa National Park; New Zealand Forest Service, 1986, n.d.).

However, the advances made in the late 1970s and the early 1980s with respect to wilderness preservation were to come to an abrupt end in 1987 with the development of new institutional arrangements for the Conservation Estate, coinciding with a period of rapid growth of international tourism.

New economic conservation (1987 – present day)

Environmental administration in New Zealand was completely overhauled by the fourth Labour Government between 1984 and 1990. Environmental reform under the Labour government was initiated, by the Environment Act 1986 (Palmer, 1990). The enactment of this legislation established the Ministry for the Environment (MfE) and the position of Parliamentary Commissioner for the Environment, described by Palmer (1990) as 'an independent guardian to protect the environment'. The third major reform to environmental administration in New Zealand was the creation of the Department of Conservation, the organization responsible for the Conservation Estate, which came into existence at the start of the financial year, 1 April, 1987.

The Department of Conservation was the result of a complete restructuring of the amalgamated parent bodies of the Department of Lands and Survey and the New Zealand Forest Service. DOC was established on a four-tier structure consisting of Head Office (Wellington), eight regional offices, 34 district offices and numerous sub-offices. The Department of Conservation became charged primarily with the management of the Conservation Estate including all national parks and was obliged to foster tourism and recreational use of heritage resources (Cahn and Cahn, 1989). In addition, the management of reserves, forest parks and other state forests, wildlife and native plants, historic foreshores, seabeds, lakes and rivers, marine resources and marine mammals were drawn together under the DOC umbrella (Crabtree, 1989; Molloy, 1993).

In 1987 New Zealand also began a decade of intense growth in international tourism arrivals. Over the decade (1987–1997) overseas visitors assumed an increasingly significant visitor presence in the national parks. In 1990 the New Zealand Tourism Board (NZTB) received $40 million of government funding in order to promote New Zealand to carefully selected tourist-generating markets. This figure was increased to $55 million in 1991 confirming the seriousness with which tourism growth targets were being pursued. By 1991 international visitors represented 65% of all users on New Zealand's 11 most popular back-country tracks; 85% of which was concentrated on only five tracks: the Abel Tasman, Milford, Routeburn, Kepler and Lake Waikaremoana Tracks (Duncan and Davison, 1991). In the year to April 1992 New Zealand received one million international arrivals for a 12-month period for the first time. The fact that this was regarded by the NZTB as merely a stepping stone *en route* to a goal of three million annual visitors by 2000 was received with some disbelief by conservationists (Higham, 1996). Due to a combination of an over-ambitious target and the impact of the Asian financial crisis on inbound tourism, the figure for the year 2000 will be substantially less than two million arrivals. However, New Zealand is still pursuing a growth strategy with respect to tourism (Hall *et al.*, 1997).

Despite DOC's key role with respect to tourism management in New Zealand, its government funding fortunes stand in stark contrast to that of the NZTB. Since 1987 DOC has suffered a 20% funding decrease in real terms. In 1995 funding for DOC's Recreation, Facility and Visitor Services was cut substantially. Only after the deaths of 14 people at Cave Creek (Westland Convervancy), when an inadequately designed and constructed viewing platform collapsed did the Government react. A one-off emergency fund to upgrade dangerous facilities was widely recognized as a reactive and inadequate measure. In 1996 a restructuring of DOC resulted in the abolition of the Recreational Services Division which was responsible for tourism planning and management. The current National Party government, which first came into power in 1991, remains firmly entrenched in its view that DOC should generate 30% of its operating costs. The result has been lengthy public debate on the merits of options for the generation of income by the Department. One such option is the levelling of access charges or charges for overnight facilities. The former has received vehement opposition from the New Zealand public which sees free and uninhibited access to the national parks as a corner-stone of New Zealand's back-country culture. Calls have also been made for international visitors to make payment for use of the Conservation Estate. For example, a conservation group, the Forest and Bird Society, has proposed to levy tourists on arrival at the international gateways. A green levy of $20.00 per tourist upon arrival in New Zealand would generate approximately $NZ30 million per annum for conservation. This, however, was opposed by the NZTB and the Ministry of Commerce's Tourism Policy Group (now the Office of Tourism and Sport) who argued that tourists bring $4.5 billion into the economy already and a charge at the point of entry would weaken New Zealand's competitive appeal. Furthermore, such a charge was considered unfair when tourist use of the Conservation Estate varies between individual tourists.

Most recently, the Business Round Table, an influential business interest group, produced a report titled 'Conservation Strategies for New Zealand', which proposes privatization of parts of the Conservation Estate. This would inevitably result in charges for entry to the Conservation Estate. While this is also seen as a means of generating income, the Hon. Nick Smith, Minister for Conservation, states that it fundamentally contradicts the core concept of the National Parks Act, free public access. More likely is the continuing issuing of concessions for entrepreneurs to operate guided walks and other operations with a percentage of profits to DOC. Regardless of these options, it is clear that DOC has been forced to dwell at length on financial rather than conservation issues. The consequence has been the unenviable task of ascribing priorities to conservation goals, including wilderness management.

The 'new economic conservation' philosophy has been pursued despite concerns over the capacities of sites within the Conservation Estate to cope with projected tourist demand (DOC/NZTB, 1993). DOC/NZTB (1993) identified that levels of use at many forested sites such as the Routeburn, Milford, Rees-Dart, Dusky and Copland tracks as well as Great Barrier Island were at or beyond their capacities to cope. Some tracks were considered capable of receiving increased use. These included the Abel Tasman, Kepler, Greenstone and Stewart Island tracks. Others, including some of New Zealand's most significant wilderness tracks, the Caples, Wilkin-Young and Hollyford, were considered capable of accommodating double the levels of use that they were then receiving, in order to meet projected levels of demand. These scenarios have been criticized on the grounds that they seek to maximize levels of visitor use rather than managing the qualities of experience that different sites offer (Higham, 1996).

Concern for the management of forested wilderness areas in New Zealand also extends to areas on the fringe of the Conservation Estate. Various issues have caused public concern for tourism and non-tourism activities in areas that buffer the Conservation Estate. The formation of the Backcountry Skier Alliance (BSA) as an Incorporate Society in 1991 is an illustration of concern for the undermining of wilderness recreational opportunities in natural areas that do not fall within the Conservation Estate. This group was created with the objectives of fostering non-motorized winter recreation in back-country areas and to promote and protect resources for winter wilderness recreation. Recent efforts of the BSA have centred on the development and expansion of commercial motorized Skidoo and Snowcat operations (and the development of infrastructure including airstrips for fixed wing aircraft access) in and on the fringes of the proposed, but as yet ungazetted, Garvies Winter Wilderness Area.

The tourism and recreation values of fringe forest resources are also threatened by extensive logging in various native forests in New Zealand's South Island. The Maruia Society identifies a loophole in the Forests Act that allows private landholders to clearfell extensive native beech forests. In 1993 the National Government passed an amendment of the Forests Act requiring sustainable management of native forest resources held in private hands. However, 400 land blocks returned to Maori in 1906 (under the South Island Landless Natives Act) in reparation for previous land claim injustices (Gibb, 1999) were exempted from this legislation. During this year alone blocks of native forest in Tuatapere, the Rowallan Burn catchment and on Humps Ridge, fringe areas that buffer Fiordland National Park and the Te Waipounamu World Heritage Area, have been logged. The same loophole is allowing the unsustainable logging of the fringes of the Catlins State Forest (south-east South Island) which provides the scenic centrepiece of the region's tourism industry. To date the government has negotiated successfully with Maori regarding only one of these 400 land blocks for the protection of native forests. One of New Zealand's leading conservation groups, the

Maruia Society estimates that a further NZ$80 million will be required in compensation if these areas are to be brought under the full jurisdiction of the Forests Act.

Contemporary Solutions: Approaches to the Management of Wilderness and Forest Areas

Kliskey (1992) identified two main approaches to wilderness management in New Zealand. Firstly, wilderness management may be approached from the point of view of maintaining its natural character, under the guidelines of the Wilderness Policy, through the designation of buffer zones and remoteness of access. This approach promotes the importance of managing buffers as an indirect form of wilderness management. The second, according to Kliskey, conceptualizes wilderness as one extreme of a recreation opportunity spectrum, servicing the needs of 'remoteness seekers' to draw directly from the DOC Visitor Strategy (1995). Both approaches give rise to the management of wilderness in terms of recreational experiences as opposed to the 'intrinsic qualities of wilderness that give rise to that experience' (Kliskey, 1992: 69). These approaches to wilderness management are a legacy of the management era that preceded the creation of DOC in 1987 as DOC's parent agencies considered that the act of designating wilderness areas was sufficient management of the conditions therein. However, the philosophy that designation is equal to protection is extremely problematic in terms of effective environmental management (e.g. see Godin and Leonard, 1979; Lucas, 1982; Haas *et al.*, 1987). In the New Zealand context, research suggests that application of the term wilderness to designated natural areas may actually encourage visitation (Kearsley, 1983; Kliskey 1992; Molloy, 1993; Booth and Cullen, 1995), which is likely to cause unacceptable impact if not carefully managed. Given the pressure of visitor numbers (Cessford and Dingwall, 1996) the time has long passed when a wilderness area could be legislatively designated and then expected to be able to maintain its wilderness qualities by virtue of its size and terrain. Despite this, Kilskey (1992: 69) employed the term '*de facto* wilderness' to describe extensive areas in New Zealand that are effectively unmanaged.

Throughout much of the 1990s DOC has approached wilderness management as part of a wider recreation resource management process within the Conservation Estate. The Recreation Opportunity Spectrum (ROS) management framework has been adopted by DOC, for which a seven-fold user classification system[1] has been developed. Therefore, wilderness has become one component part within a planning process that provides for a range of recreational experiences. This management system can been criticized on two grounds. Firstly, it fails to recognize that for most people wilderness experiences can be achieved in a range of natural settings that may or may not be designated as wilderness (see Chapter 5). The use of the term wilderness within the DOC ROS classification fails to observe this point. Secondly, ROS fails to incorporate ecological conditions into the wilderness management process in the manner that planning frameworks such as Limits of Acceptable Change (LAC) and Ultimate Environmental Thresholds (UET) are designed to achieve (Hall and McArthur, 1998).

Redefining wilderness in New Zealand

The management of wilderness at one extreme of the Recreation Opportunity Spectrum has generated substantial management difficulties for DOC. These can, once again, be linked to the historical development of the wilderness system and the legislation governing its management. DOC manages Wilderness Areas in New Zealand in accordance with the Wilderness Policy (outlined above) created by the Wilderness Advisory Group, both products of the Wilderness Conference hosted in 1981 by the Federated Mountain Club. Their Wilderness Policy was adopted by the Minister of Lands and Forests in 1983 and endorsed by the Minister of Conservation for DOC in 1989.

As outlined above, this policy requires that wilderness be managed to maintain its natural state, without buildings of any description, animals or vehicles, roads, tracks or trails. Even in 1981 when ten wilderness areas were proposed by the FMC under these conditions, it was apparent that such wilderness designations would be both relatively few and relatively small in size (Table 9.1). In other words, the criteria set forth for the gazettal of wilderness areas in New Zealand were so purist that few remaining areas met the standards required of wilderness. The outcome of this has been a spartan and scattered wilderness system. A number of proposed wilderness areas have not been gazetted due to the setting of alternative conservation priorities for the Department. Some remain outside the Conservation Estate while three existing wilderness areas (Otehake, Te Tatau-Pounamu and Hauhungatahi), each gazetted prior to the Wilderness Policy, fail to meet the criteria of Wilderness Areas on the grounds of size (these areas range from only 6500–12,000 ha).

Consequently, according to definition, the wilderness pole of the Departments ROS classification is represented by small pockets of designated wilderness. Indeed, as a response to this, DOC (1995) has outlined a series of management principles which, whilst in accordance with the 1985 Wilderness Policy, indicated an easing of the purist approach to wilderness designation and management. These would indicate that a new set of institutional arrangements for wilderness management have been put in place. The following extracts from the 1995 DOC Visitor Strategy confirm a shifting of wilderness management principles (italics added for emphasis):

> Wilderness areas will be managed:
>
> To retain natural wilderness qualities, developments such as huts, tracks, route markers and bridges are inappropriate, *and in the few cases where such facilities exist they should be removed or no longer maintained.*
>
> Adjoining lands should be managed as buffers to assist in the protection of a wilderness area; buffers may contain huts, tracks and bridges, but these should be few and *vehicle access will be discouraged near the wilderness boundary.*
>
> To ensure the use of wilderness areas at levels compatible with the maintenance of wilderness values, *commercial recreation activities may only be undertaken under licence or permit.*
>
> Because wilderness areas are places for quiet enjoyment, free from obvious human impact and require physical endeavour to achieve in full measure the wilderness experience, the use of powered vehicles, boats or aircraft will not be permitted; *the use of horses may be allowed where strong historical links exists and where legislation permits.*
>
> Because of the overriding importance of protection of intrinsic natural values and the safety of visitors to wilderness areas; *restrictions on air access may be lifted temporarily for management purposes such as search and rescue operations, fire fighting, and control of introduced plants and animals*
>
> (Source: DOC Visitor Strategy (1995))

These statements leave little doubt that the WAG's (1985) Wilderness Policy has been diluted in terms of actually managing wilderness areas. This revised wilderness management strategy may ultimately facilitate an expanded wilderness area system with greater access for visitors. This may be at the expense of the qualities which were originally the focus of the system. This may contribute to the inclusion of areas into the wilderness system that are somewhat removed from the wilderness qualities required to be designated as wilderness under the WAG Policy. These policies clearly amount to a significant downgrading of the WAG Wilderness Policy (1985). The implications of the Visitor Strategy for the management of areas that already exist within the wilderness system is open to question.

Conclusions

Environmental management strategies are not unchanging. The institutional arrangements which surround the management of the forest wilderness in New Zealand have shifted over time. More significantly, this has meant that the environmental management strategies that are utilized with respect to tourism have also changed. This chapter has argued that the environmental management settings for New Zealand's forest wilderness areas are greatly determined by the institutional arrangements in which they are set. While much research on the use of New Zealand's forest wilderness has focused on the perceptions of individual visitors (Higham, 1996; see also Chapter 5) the organization of the supply of such tourist experiences has been given relatively little consideration. However, it is the institutional arrangements which ultimately provide for forest wilderness experiences.

An examination of the institutional arrangements for the provision of forest wilderness experiences in New Zealand suggest that institutional change, usually tied into the development of new legislative structures, followed by periods of relative stability, is the norm. Present policy settings for the provision of wilderness experiences in designated areas appear to be changing in relation to demands for increased access for international and domestic visitors. Such shifts may have substantial consequences for the forest wilderness itself because wilderness experiences are reliant on the provision of areas with high quality primitiveness and remoteness (Hall, 1992). Therefore, in New Zealand, as with many other natural areas around the world, difficult decisions regarding designation of wilderness areas and reserves and rights of access need to be made if the resource base is not to be further impaired. However, while government and tourism organizations such as the NZTB continue to focus on encouraging visitation insufficient attention is being given to maintaining the resource base. Sustainable forest management practice requires that attention be given to both the demand and the supply side of tourism. By maintaining the historical focus on the visitor rather than the resource, the present economic emphasis serves to reinforce the original designation of natural areas as 'useless' land which only gains value through tourism, rather than the intrinsic value of the resource itself.

Notes

[1] The visitor classification system developed by the Department of Conservation includes: Short stop travellers, Day visitors, Over-nighters, Back-country comfort-seekers, Back-country adventurers, Remoteness seekers and Thrill seekers.

References

Boas, A. (1993) Tourism. Letter to the Editor, *Otago Daily Times.* 15 September, 8.

Booth, K. and Cullen, R. (1995) 'Recreation impacts'. In: *Outdoor Recreation in New Zealand,* Vol 1. *A Review and Synthesis of the Research Literature.* Department of Conservation and Lincoln University, Wellington.

Cahn, R. and Cahn, P.L. (1989) Reorganising conservation efforts in New Zealand. *Environment* 31 18–20, 40–45.

Cessford, G.R. and Dingwall, P.R. (eds) (1996) *Impacts of Visitors on Natural and Historic Resources of Conservation Significance. Part 1 – Workshop Proceedings.* Science and Research Internal Report, No. 156. Department of Conservation, Wellington.

Crabtree, P.S.J. (1989) A nature conservancy for New Zealand: The Department of Conservation – its genesis. Unpublished essay submitted in partial fulfilment of the Postgraduate diploma in Political Studies, University of Otago, Dunedin.

Department of Conservation and New Zealand Tourism Board (1993) *New Zealand Conservation Estate and International Visitors.* Department of Conservation and New Zealand Tourism Board, Wellington.

Department of Conservation (1995) *Visitor Strategy*. Department of Conservation, Wellington.

Department of Lands and Surveys (1984) Report of the Department of Lands and Surveys for the Year Ended 31 March 1984. Government Printer, Wellington.

Duncan, J. and Davison, J. (1991) *Review of the Capacity of Selected Tramping Tracks to Cater for Projected Increases in Overseas Trampers.* NZTB, Wellington.

Federated Mountain Clubs of New Zealand (1981) Statement of conclusions on wilderness endorsed by full conference. *50th Jubilee Wilderness Conference, 22–24 August, 1981*, Federated Mountain Clubs of New Zealand, Wellington.

Fleming, C. (1979) The history and future of the preservation ethic. In: *National Parks of New Zealand, Proceedings of the Silver Jubilee Conference of the National Parks Authority of New Zealand, Lincoln College, 5–8 July, 1978*, National Park Series No. 14, National Parks Authority of New Zealand, Department of Lands and Survey, Wellington, pp. 54–64.

Gibb, J. (1999) Catlins' tourism potential fears. *Otago Daily Times*, Wednesday 7 April, 3.

Godin, V.B. and Leonard, R.E. (1979) Management problems in designated wilderness areas. *Journal of Soil and Water Conservation* 34, 141–143.

Government Caucus Committee (1979) *Review of the Administrative Structure of National Parks and Reserves Administered by the Department of Lands and Survey, July*. Government Caucus Committee, Wellington.

Haas, G.E., Driver, B.L., Brown, P.J. and Lucas, R.G. (1987) Wilderness management zoning. *Journal of Forestry* 85, 17–21.

Hall, C.M. (1988a) The geography of hope: the identification and conservation of Australia's wilderness. PhD thesis, Department of Geography, University of Western Australia, Nedlands.

Hall, C.M. (1988b) Wilderness in New Zealand. *Alternatives: Perspectives on Science, Technology and the Environment* 15, 40–46.

Hall, C.M. (1992) *Wasteland to World Heritage: Wilderness Preservation in Australia.* Melbourne University Press, Carlton.

Hall, C.M. (1993) John Muir's travels in Australasia 1903–1904: their significance for environmental and conservation thought. In: Miller, S. (ed.) *John Muir: Life and Work.* University of New Mexico Press, Albuquerque.

Hall, C.M. (1998) Historical antecedents of sustainable development and ecotourism: new labels on old bottles? In: Hall, C.M. and Lew, A. (eds) *Sustainable Tourism Development: Geographical Perspectives.* Addison-Wesley Longman, Harlow.

Hall, C.M. and Jenkins, J. (1995) *Tourism and Public Policy*. Routledge, London.

Hall, C.M., Jenkins, J. and Kearsley, G. (eds) (1997) *Tourism Planning and Policy in Australia and New Zealand: Issues and Cases.* Irwin Publishers, Sydney.

Hall, C.M. and McArthur, S. (1998) *Integrated Heritage Management.* Stationery Office, London.

Hall, C.M. and Shultis, J. (1991) Railways, tourism and worthless lands: the establishment of national parks in Australia, Canada, New Zealand and the United States. *Australian-Canadian Studies – A Journal for the Humanities & the Social Sciences* 8, 57–74.

Harris, W.W. (1974) Three parks: an analysis of the origins and evolution of the national parks movement. Unpublished MA thesis, Department of Geography, University of Canterbury, Christchurch.

Higham, J.E.S. (1996) Wilderness perceptions of international visitors to New Zealand: the perceptual approach to the management of international tourists visiting Wilderness areas within New Zealand's Conservation Estate. Unpublished PhD thesis, University of Otago, Dunedin.

Kearsley, G.W. (1983) Images of wilderness and their implications for national park management. *International Imagery Bulletin* 1, 32–38.

Kearsley, G.W. (1997) Managing the consequences of overuse by tourists of New Zealand's Conservation Estate. In: Hall, C.M., Jenkins, J. and Kearsley, G. (eds) *Tourism Planning and Policy: Cases, Issues and Practice.* Irwin Publishers, Sydney.

Kearsley, G.W. and Higham, J.E.S. (1997) Management of the environmental effects associated with the tourism sector. Review of literature on environmental effects. Report prepared for the Parliamentary Commissioner for the Environment, Wellington.

Kelly, G.C. (1980) Conservation – a biologist's viewpoint. In: Bishop, W.J.F. (ed.) *Preservation and Recreation.* Land Use Series No. 7, Department of Lands and Survey, Wellington.

Kliskey, A.D. (1992) Wilderness perception mapping: a GIS approach to the application of wilderness perceptions to protected areas management in New Zealand. Unpublished PhD thesis, University of Otago, Dunedin.

Lucas, P.H.C. (1970) Conserving New Zealand's heritage: report on a study tour of national park and allied areas in Canada and the United States. Government Printer, Wellington.

Lucas, R.D. (1982) Recreation regulations: when are they needed? *Journal of Forestry* 80, 148–151.

Mitchell, B. (1989) *Geography and Resource Analysis.* Longman Scientific and Technical, Harlow.

Molloy, L.F. (1993) The interpretation of New Zealand's natural heritage. In: Hall, C.M. and McArthur, S. (eds) *Heritage Management in New Zealand and Australia: Visitor Management, Interpretation and Marketing.* Oxford University Press, Auckland.

National Parks Authority (1978) New Zealand's National parks: National Parks Authority General Policy. National Parks Series No. 9, Department of Lands and Survey for the National Parks Authority, Wellington.

New Zealand (1887) *Parliamentary Debates* 57, 399.

New Zealand (1894) *Parliamentary Debates* 86, 579.

New Zealand Forest Service (1986) Proposed Tasman and Paparoa Wilderness Areas, Document No. FS 36/4/1 6/0/12 (24245E), New Zealand Forest Service, Wellington, 7/11/86.

New Zealand Forest Service (undated, 1986?) Recommendations for the Proposed Raukumara Wilderness Area, Document No. FS 6/0/12 (1615E), New Zealand Forest Service, Wellington.

O'Neill, D. (1994) Socially sustainable tourism development. Unpublished MA (Tourism) thesis, University of Otago, Dunedin.

O'Riordan, T. (1971) *Perspectives on Resource Management.* Pion, London.

Ostrom, E. (1986) An agenda for the study of institutions. *Public Choice* 48, 3–25.

Palmer, G. (1990) *Environmental Politics. A Greenprint for New Zealand.* John McIndoe, Dunedin.

Runte, A. (1972) Yellowstone: it's useless, so why not a park. *National Parks and Conservation Magazine,* 46 (March), 4–7.

Runte, A. (1973) "Worthless" lands – our national parks: the enigmatic past and uncertain future of America's scenic wonderlands. *American West* 10, 4–11.

Runte, A. (1979) *National Parks: The American Experience.* University of Nebraska Press, Lincoln.

Scrutton, R. (1982) *A Dictionary of Political Thought.* Pan Books, London.

Thomas, A.P. (1891) The presentation of the native fauna and flora of New Zealand. *New Zealand Journal of Science* 2, 93.

Thompson, J. (1976) *Origin of the 1952 National Parks Act.* Department of Lands and Surveys, Wellington.

Veblen, T.T. and Stewart, G.H. (1982) The effects of introduced wild animals on New Zealand forests. *Annals of the Association of American Geographers* 72, 372–397.

Wilderness Advisory Group (1985) *Wilderness Policy.* Department of Lands and Survey and New Zealand Forest Service, Wellington.

Wynn, G. (1977) Conservation and society in Nineteenth Century New Zealand. *New Zealand Journal of History* 11, 124–136.

10

From Wasteland to Woodland to 'Little Switzerland': Environmental and Recreational Management in Place, Culture and Time

Paul Cloke and Owain Jones

> One of the basic premises of woodland ecology, [is] that woods are more than just collections of trees. They are *places*, landmarks and communities.
>
> (Mabey, 1980: 63, emphasis as original)

Introduction

Our aim in this chapter is to apply to a specific site some of the more recent conceptualizations of 'arbori-culture' – that is the cultural constructions of the 'place' of trees, and the significance of trees in the construction of their 'place'. This work has its roots in developments within human geography's concern for landscape and place, in broader considerations of nature–society relations in social theory, and in environmental philosophy. This chapter draws upon our wider research[1] which seeks to understand how places of trees come to be formed and imagined, and which links back to the practical concerns of those responsible for the management of such sites, be they parks, woodlands, forests, orchards, or other places where trees are major contributors to the place milieu. In particular we feel that the story of a particular site which is told below, highlights the complexities, not to say difficulties, with processes of 'stakeholder involvement' which now abound in the culture of governance (see Rhodes, 1997). The culture of partnerships which include state, private sector, voluntary sector and 'the community', often bidding for 'challenge funding', which marks this recent configuration of governance, is now a feature of many areas of public function, including environmental management. The seemingly virtuous nature of such processes in fact conceals a whole host of questions which have come under

(eds Xavier Font and John Tribe)

critical scrutiny (see for example Peck and Tickell, 1994), and now not least in the context of local rural state function (Goodwin, 1998; Murdoch and Abram, 1998; Cloke *et al.*, 2000). In terms of environmental management systems and how these might be applied to woodland management, it is equally important to recognize that 'stakeholder involvement' is a complex notion fraught with all sort of unpredictable issues. These will include local knowledge (contemporary and historic), values, power relations, contested community identities, and the contingent way in which all the components – cultural, ecological, political, economic and so forth – of a *particular place* generate a unique 'chemistry' which needs to be accounted for.

In what follows we first briefly outline in more detail the theoretical trajectories which are driving our approach – place, culture and agency. Then we rework these ideas in the context of a particular tree-place in order to show how looking through these lenses may provide a conceptual frame with which to underpin understandings of, and aspects of the environmental management of 'places of trees'.

Place

Place is a fundamental aspect of existence, we (humans and non-humans) are all in place, in some way or other, at all times (Casey, 1998), yet the term 'place' is one with the greatest range of possible meanings (Harvey, 1996). It is at once a very obvious and very illusive notion. Without getting drawn into ontological quagmires here, we want to consider place in ways which have been vividly articulated by the environmental organization, Common Ground. Their notion of place revolves around the idea of 'local distinctiveness', suggesting place to be some form of physical/imaginative space, be it, perhaps, a village, farm, park, wood, forest or region, which is, in some way, identified as having some internal cohesion distinct from that around it. Such distinctions may be material or cultural, and will usually be a complex construction of differing elements. There may well be overlapping scales of distinctiveness, and any achieved place-identity will always be subject to contestation, change, partiality, fading and reforming, and also complex interconnections between the global and the local (see Massey and Jess, 1995). Drawing on the work of anthropologist Tim Ingold (1993), we see this notion of place as a manifestation of 'dwelling', where all manner of elements – people, artefacts, animals, plants, topography, climate, culture, economy and history – are knotted together in an utterly unique way to form unfolding space–times of particular landscapes and places. Such an approach overcomes many of the epistemological weaknesses which have beset academic approaches to nature, the environment and landscape. For example Macnaghten and Urry (1998) suggest that nature, seen as landscape which is conceptualized from a dwelling perspective, offers exciting new perspectives which contrast the incomplete narrowness of three dominant views of the environment – 'realism', 'idealism' and 'instrumentalism' – and also the great divide of nature and culture which still dominates many views of the world. This approach offers a way to deal with the 'richness' of place, where the ecological and the cultural, the human and non-human, the local and the global, the real and the imaginary all grow together into particular formations in particular places. It is this very richness which Common Ground has been attempting to defend from the homogenizing tendencies of industrialization, modernization and globalization.

Culture

One set of key, ever-present, but difficult to track, elements of such place milieu as outlined above, are cultural formations. These are common understandings which may be at large in society at a number of scales (national, regional, local) and locations (media, local/national state), which come to contribute to the specific matrix of differing place milieu. Places are, in part, outcomes of the coming together of local, regional and national cultural constructions, in co-present material, social, economic and historical contexts. Cultural attributes are often particularly significant in the consideration of places where trees are characteristic of place-milieu. For example Schama (1995) tells of differing national-scaled cultural inscriptions of tree meanings, which influence the complex ways in which tree-places are constructed locally. Trees in the UK and elsewhere have become carriers of peoples' environmental anxiety and love for nature, cropping up in various discourses on environmental crisis, countryside change and habitat loss, and quality of urban life. More generally trees have long been symbols for all manner of key social meanings and practices (Rival, 1998), for example being associated with fear, spirituality, recreation and so forth (Cloke *et al.*, 1996). Within these broad, but often very powerful understandings of trees, more specific variations will occur, for example the understanding of trees as native or alien; the associations attached to groups of trees such as evergreen and deciduous; wild or planted; the concern for ancient trees; and the associations with differing types of trees such as the oak in English culture.

This coming together of local and national cultural constructions, with the particular presences and juxtapositions of elements making up particular places makes for an extremely fine-grained set of relations, meanings and significances, which provide both potential opportunities and problems for environmental management of any type of site, and particularly where trees are involved, given the huge amount of cultural, emotional and even spiritual baggage certain trees in certain places can carry. To illustrate this, one of the inspirations for our research was the case of a group of lime trees in Bristol (UK), in which, as in one of the many high profile tree protest sagas of the early 1990s, prolonged campaigning, including a lengthy tree-sitting, failed to prevent a number of mature trees being felled to make way for access to a new Tesco Store at Golden Hill. These trees, although fine specimens, were not of *particular significance* through the lenses of historic landscape, ecological richness or scientific interest, so they fell outside, or perhaps through the nets of conservation legislation. Yet in the context of their place, they were extremely significant to local people, and for objectors who came from further afield, they were in the front-line of institutional destruction of nature. Place presents difficulties for legislative practice because by its nature legislation is designed to take a standardized overview of practice, and to apply blanket definitions of values, which inevitably miss the fluctuation in values which might occur through configurations of place. It should not be forgotten that these fluctuations might be sufficiently extreme to negate the generally positive understanding of trees, and for example make places of trees places of fear (Burgess, 1993), thus producing local endorsement of tree felling or thinning which might be in conflict with management objectives.

Non-human Agency

Alongside these notions of place milieu and cultural constructions we want to introduce one more theme, that is the agency which the trees themselves bring to the ongoing unfolding of places and landscapes. In understandings of human–nature relations it is now recognized that nature is not merely inscribed upon by human culture and human practice, but rather 'pushes back' with its own vitality which is manifest in specific material processes. Whatmore and Boucher (1992: 167/8) in their analysis of social constructions of nature stress that 'while nature cannot be (re)produced outside social relations, neither it is reducible to them. Rather, the biological and physical dynamics of life forms and processes need to be recognized on their own terms, conceptually independent of human social agency, such that social nature represents "the outcome of a specific structure of natural/social articulation" '. In this way, agents of nature are now seen as palpably active, not only in terms of their own biological sense, but also when bound up in the construction of social, economic, cultural, political and material formations. Harvey (1996) deploys the term 'socio-ecological' processes to encompass this fundamental stance, and stresses the need to consider non-human agency within it. And given that Harvey (and others) call for a dis-aggregation of the homogenized term 'nature', into its various 'intensely internally variegated [,] unparalleled field of difference' (Harvey, 1996: 183), the agency of, in this case trees, and of differing kinds of trees, needs to be taken seriously. We are aware that to talk of trees having agency is to invite scepticism, or worse. However, trees palpably are active, and are active in ways which are purposive (as a fulfilment of their embedded tendencies to grow in certain ways and reproduce), transformative and even creative. These qualities are constituent parts of agency (Mele, 1997), and once we try stopping to squeeze all notions of agency through the very human grid of language (and thought;

Callon and Law, 1995), the capacity for agency can be redistributed throughout a heterogeneous set of actors, including non-human actors. (These issues in relation to trees are explored more fully in Cloke and Jones, 1999.) Such understandings have particular relevance to philosophical debates within social theory (Latour, 1993), human geography (Whatmore, 1999), and environmentalism (Plumwood, 1993), but also we suggest to issues of environmental management. When any site is considered as a milieu of physical and cultural elements, trees will play an active role within this process, projecting themselves into political, cultural and economic fabrics, and through the historical geographies of these fabrics as articulated in the changing nature of places and landscapes. This is evident in Watkins' (1998a) study of the ancient oaks in Sherwood Forest which 'can be so long-lived that they develop several layers of meaning that can be documented through time' (Watkins, 1998b: 8), and it is clear that such differing meanings contribute to shifting constructions of the Forest as a place over time.

We therefore suggest that 'places of trees' be they forests, orchards, woods, parks, need to be understood as the result of the 'dialectical unfolding of historical and ecological processes' (Rival, 1998: 24) and this is so even of trees and forests usually considered to be 'naturally given categories'. Watkins (1998a: 1) suggests that 'the history of European woods and forests have long remained somewhat on the edge of academic study' and later concludes that the emerging approaches and studies he considers 'contribute to an increasing *uncertainty* about woodland and forest history', because they show the particularity of each case, and bring ecological, historical, and cultural specificity to what were before mainly quite broad categorizations.

> It is increasingly difficult to accept woodland as a simple category from which a settled landscape is wrought. Rather it must be seen as a complex type of land use which has varied dramatically in the density, age, species and forms of trees and shrubs of which it consists. The utility of woodland and the cultural values ascribed to it are also diverse ... To explore this complex bundle of uses and values is no simple matter.
>
> (Watkins, 1998a: 1)

We suggest that using the concepts and categories we have outlined provides a conceptual framework for approaching treed landscapes which takes account of these important aspects of localized specificity.

Camerton

We now narrate the story of a site which, told at its briefest, was a bare, working coal mine spoil heap at the beginning of this century, which became wooded over time, was acquired by the local community as an open space in 1987, was subsequently developed and managed as a mining and natural heritage site, and which in 1997 was designated as a local nature reserve.

Due to the presence of coniferous trees which were planted on 'the old batch' early this century ('batch' is the local name for coal spoil tips), and other trees which had self-seeded and grown on the site, it has been viewed by the local community, and subsequently, managed, funded and argued over, as a woodland site. In what follows we consider the particular unfolding 'socio-ecological' construction of this site through the conceptual lenses set out above. We suggest that all tree-places will be formed in part through such processes, and that the approach we take offers the opportunity to understand both the specificity of emplaced nature, and how such places are constructed culturally and naturally.

Earlier historical background

Camerton was one of the 70 or so collieries which comprised the Somerset Coalfield in south-west England, the last of which closed in 1973. Although there was evidence of mining in mediaeval times and before, mining in the area burgeoned in relation to the industrial revolution, and Down and Warrington (1971) make 1790 the beginning of this modern period of coal extraction which continued until a decline began at the turn of this century. The Camerton colliery, named as such due its location in the village and parish of Camerton (now in the county of Bath and North East Somerset, UK), consisted of two pits, the old pit being sunk in the 1780s, and the new pit being operational by 1800. The old pit was closed for coal winding in 1898, but the shaft was kept open for ventilation and access for the new pit. When the new pit was connected underground to a neighbouring pit (c. 1930) the old pit was closed and dismantled. The new pit continued to produce coal until closure in 1950. Thus both pits were producing coal, and colliery spoil, for a century or more, and like other pits, the spoil was tipped close to the pithead and gradually the huge spoil heaps characteristic of coal mining areas were formed.

Figure 10.1 shows a map of the Camerton colliery in 1883 with the spoil heaps spreading out from the pitheads of both old and new pits. At this time the technique used to move the spoil was horse- or man-drawn wagons running on rails. The tips were slowly built up with the track being extended as the tip expanded. If topography allowed, the spoil would be tipped below the level of the pithead, thus avoiding having to drag the spoil upward. At Camerton the ridge of the old tip which marks the route of the extending track rises up gradually, thus gaining height which increases the capacity of the tip site. As the ridge of the tip gradually rises, and the ground below drops away, the sides of the tip become increasingly high and steep. This new topography provided a distinctive form which was inherited and adapted by the subsequent uses of the site. In the case of Camerton, as Fig. 10.2 shows, mechanical spoil-moving devices which were capable of moving the spoil up steeper inclines were introduced around the turn of the century, thus more spoil could be deposited in the same area, and the more

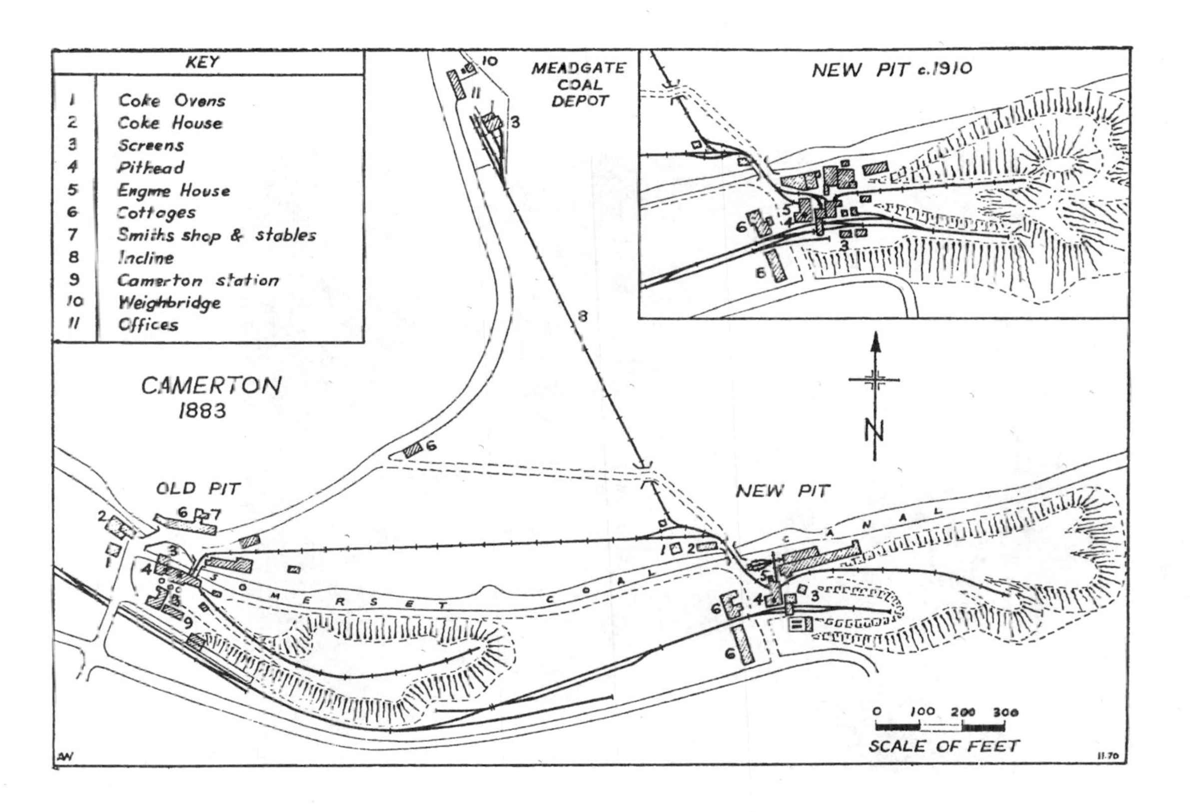

Fig. 10.1. Map of the Camerton Colliery, 1883, with insert of new pit, 1910, showing spoil heaps. (From Down and Warrington, 1971, reproduced with kind permission of the authors.)

Fig. 10.2. Photograph of the new pit batch, possibly 1930s, showing a Mclean Tipper, the steeper tip form, and the bare condition of the tips when working.

modern conical spoil heap formed (see 1910 inset, Fig. 10.1). The new batch thus consists of an early section of old type tipping with the conical spoil heap built on top of that.

Interestingly, in the Somerset coalfield, more than in many other mining areas (for example the South Wales coalfield), the coal tips have remained relatively undisturbed in terms of clearance or alteration. The reasons for this, and the consequences, have received very little study (not least as a possible related series of ecological regeneration case studies). As in other coalfields the tips became prominent features of the local landscape, and were seen, by some at least, in negative terms. Little (1969) in his portrait of Somerset wrote that, Camerton was 'still a little gaunt with the relics of its colliery, and with the great pyramid of spoil which blotches the hillside' (p. 23). One factor in the fate of these coal tips was the generally rural landscape in which they were set (they are also smaller than the vast tips which added to the more 'industrial' landscapes of South Wales). To an extent the tips were 'reabsorbed' into the rural, hilly landscape, and Buchanan and Cossons (1969) noted that since the close of the mines 'the process of reversion to rural countryside has gone on apace' and the large spoil heaps at Camerton were 'now becoming overgrown' (pp. 96, 97). However, the story of the Camerton batches and the other batches are not as straightforward as this suggests. Firstly a significant number were planted with conifers in the early part of this century, and these remain a distinguishing feature of the local landscape. Secondly the natural re-colonization which has taken place on all the batches has, according to the Cam Valley Wildlife Group (1998), added significantly to the range and diversity of habitats in the area, and given the general decline in the biodiversity of the agricultural landscape the batches now stand in, they are of considerable ecological significance. The batches, as other physical reminders of the mining days such as the canal and railway network which served the collieries were lost, have also become very significant in terms of local industrial archaeology, and particularly in terms of community histories and memories of the mining era.

The trees

Camerton 'old batch' was one of those planted with conifers. Of the 35 or so batches in the Cam Valley Wildlife Group area, half or so have areas of conifers on them, with a few being totally covered (S. Preddy, 1999, Cam Valley Wildlife Group, response to email questionnaire sent by Owain Jones). The County Council Woodland Officer who assessed the site when purchase by the local community became a possibility, and who drew up the initial management plans, recorded the conifers as a mix of Douglas fir, with some Corsican pine, Scots pine and European larch. He added that 'nobody knows exactly when they were planted but my impression was it was at the turn of the century ... certainly no later the First World War'. This planting seems to be a practice distinct to parts of the Somerset Coalfield at that time, and information about the exact timing and reasons for planting is difficult to come by. 'Local knowledge', including that of those who have charted the history of the Camerton Colliery and who have become involved in the running of the batch as a Heritage Site, suggests that the conifers were planted in order to stabilize the batches and to provide a crop of timber for working the mine in future years. There would seem to be a certain logic to this. Firstly, the stability of larger, steeper spoil heaps has long been a concern in the mining industry (NCB, 1970). Secondly, Williams (1976: 73) in his history of another nearby Somerset colliery describes in detail how round-section timber, as would come from conifers, was the key means of construction of roadway and working faces in the mines, with 'arms', 'collars' and 'lagging' forming a frame which secured the 'rippings'. He adds that the timber was usually 'home grown', on local farms and estates, and that in difficult workings 'the cost of the timber was sometimes as much as the selling price of the coal' so the incentive for the collieries to grow their own timber would have been strong. However, two other historians of the Somerset coalfield have suggested that the conifers would not have been planted for the future production of pit props (Dr C. Chillcott, and Mr J. Cornwall, 1999, personal communication). Dr Chillcott suggests that the conifers were planted around 1920 as an 'enlightened act' of the then colliery owner, in an effort to landscape the numerous bare spoil heaps of the time, this being his understanding derived from oral history accounts. [Intriguingly Condry (1974) says that at this time the newly formed Forestry Commission (1919) was beginning to plant conifer plantations to ameliorate the deforestation affected by the First World War.]

The collieries were privately owned by the landowners whose land they stood on. Thus the owners were often large estates such as The Duchy of Cornwall and the Waldergrave Estate. Camerton was owned by the Jarrett family up until 1911 when it was sold to Sir Frank Beauchamp (Down and Warrington, 1971), who according to Macmillen (1990) aimed to 'amalgamate the collieries into a more economical system of working in order to compete with the more favoured mining districts elsewhere', and who registered the first 'Somerset Collieries Ltd' in 1925. (The coming of many of the mines into single ownership may account for the planting of many of them with conifers at roughly the same time.)

Whichever of these accounts of why and when the trees were planted is more accurate, the circumstances of the mines swiftly changed, due to the evolving economic and technical aspects in coal mining, and the conditions in the local pits where the easily accessible coal was of poor quality and the working costs were relatively high (Macmillen, 1990). In the first decades of this century many of the collieries were in decline. Camerton was taken into state ownership, along with the rest of the remaining Somerset Coalfield, when the industry was nationalized in 1947. Working continued for a few years but 'the last coal was wound on the 14 April, 1950' (Macmillen, 1990: 34), and after a brief period of salvage work the colliery was closed. The pithead structures were dismantled, the shafts capped off and the branch railway line also closed.

Once areas of the batches were completed in terms of tipping, and where they had been planted, they were left relatively undisturbed. On the closure of the mine, the whole site including the unplanted areas of new batch also lay undisturbed. While the site was dormant in terms of mining or subsequent economic development, the trees, both those planted and those which had self-seeded in favourable spaces, continued to grow. A sense of the gradual but powerful transformation created by this growth can be gathered from the accounts of local residents' memories of playing on the site as children. One resident told us,

> I should think they were planted about I should say 70, 80 years ago because when I was a kiddy those trees on that batch, well I can remember some of them being planted, but some were about 2 feet high so they had been in probably three of four year by then. [ ] Oh we always played on there, it was our only means of playing [ ] well the trees then were about 6 feet high [and by the time the mine closed] oh they were good trees then [ ] very strong trees.

Another, younger resident told us of her post-war childhood memories of the batch:

> all I can remember is tall conifers and where everything was so dense that was hardly any undergrowth there and there were lots of tracks because us kids used to play there in gangs.

Firstly, in terms of the 'creative' active capacities of trees we have considered, the trees' ability to grow in such a location is significant. The photograph in Fig. 10.2 shows what the working batches and the freshly tipped spoil was like; a bare, lifeless and apparently 'unnatural' land form. Moffat and Buckley (1995) chart the problems and importance of understanding soil conditions when planting trees on disturbed ground, paying particular attention at one point to colliery spoil. Such soil they suggest may well present a number of adverse factors to tree planting and growing, and they summarize tree species and trial results of planting on spoil heaps. In the case of Camerton it is unlikely that such information was readily available at the time and there was probably a more *de facto* recognition that if planted certain trees may well grow. Today such a capacity is still deployed in the transformation of colliery waste landscapes.

> Graham Howe, head of coalfield regeneration for English Partnerships [ ] pointed across the road to where saplings were poking through the sparsely grassed surface of the former spoil tip. 'We're putting in about 100,000 fir trees to make it look presentable and hoping to attract a foreign inward investor'.
>
> (Arnot, 1999)

In these instances, of course it *is* the human actors who are the enrollers, in Actor Network Theory terms, of these processes. But without the trees' capacity to 'tolerate' such conditions the process would be impossible, so they do contribute meaningfully to what is termed *relational agency*. The other, 'wild' trees which had and were self-seeding onto the site, and which come to play an important role later in the story, could be seen as bringing a more independent creative agency to the development of the site. Trees can become enrolling actors in networks as Brown's (1997) account of the formation the Black Forest Urban Forest Project shows.

> To the local's surprise, even where there was nothing but rubble, trees began to grow and woods were springing up on ground that was considered too

contaminated to be redeveloped. Although the Black Country was collectively taken aback by this unexpected side-effect of de-industrialisation, the citizens do have a reputation for innovation, and suddenly here was the opportunity to transform the image of Tipton, Dudley, Walsall and Wolverhampton [ ] The Black Forest Urban Forest Project was born. Old railway sidings, coal and waste tips and demolished factories have been turned into tree plantations over the last five years [ ] It was discovered that the ever-present air pollution from the overcrowded M5 and M6 was considerably reduced where there were trees. The trees trapped the dust from vehicle exhausts. The forest was therefore extended for the full 16 miles of the motorway through the black country ... So successful has the urban forest become that the Department of the Environment and Transport has transformed the Black Country scheme into a National Urban Forestry Unit with 15 staff.

Secondly, beyond this initial stage of tolerating or colonizing the site, the trees kept growing even when the cultural/economic dynamic of the original intentions behind those planted faded away. The trees broke away from their initial culturally constructed identity and role (whatever it was) and through their continued presence and growth projected themselves into new cultural constructions which came to form around them. This we suggest is a common feature of tree agency. They continue to grow and develop in the 'vacuums' which sometime appear between the breakdown of one formation of a culturally constructed landscape and the forming of a subsequent landscape configuration. As with the oaks of Sherwood Forest (Watkins, 1998b) the trees 'outlive' the social constructions which they are initially bound up in and become embroiled in new associations. Given the longevity of trees and how their agency is expressed in a differing, slower, time-frame to that of human agency, they push their developing presence through subsequent 'depositions' of place and landscape identity construction.

In this case the trees had a presence associated with the working mine, the closed mine, and the 'dormant' site. These phases spanned a period from the first decades to well into the latter half of this century, by which time attitudes to landscape, nature conservation, the environment, public access to open space, forestry and woodland management had dramatically transformed. This was not least through growing awareness of environmental and nature conservation issues at the global, national and local scale.

Other local developments had also occurred. The village of Camerton had shrunk back in population size from a mining community of some 2386 to a much smaller population of around 530. From being a industrial landscape with accompanying social factors famously portrayed by the Reverend Skinner (1971)[2] it was 'reverting' to a more traditional rural formation. As a social part of this reconstruction there was a move to enter the 'Best Kept Village' competition, and it was felt that the old pithead site (see Fig. 10.1), which was in a key position opposite the then village post office, was a problem because at this time it was overgrown with brambles. So in the early 1980s the Parish Council opened negotiations with the National

Coal Board with the aim of acquiring the pithead site, or access to it, as a public space for the village. Initially, owning and managing the batch was not part of their intentions.

Also at this time, clearing coal tips within the major coalfields of the UK was seen as a priority in terms of economic restructuring, and also marked the tail-end of the response to the Aberfan disaster. Some companies were recycling spoil heaps (recovering coal which could be used in coal-fired power stations, and aggregates for other purposes, and freeing up the recovered land for development). One company which had cleared tips in South Wales proposed to clear the Camerton batches and began discussions with the Coal Board. The current Chair of the Camerton Heritage Committee (which was formed by the Parish Council in 1989 in order to manage the site once it was eventually acquired) told us that at this point a public meeting was held about the batches and 'that's when all hell was let loose'. A resident also recalls 'there was quite an uproar ... the batch as such had to remain, locals didn't want it disturbed or anything like that, no way'. Not only was there a strong sentimental affection for the batches, there was also considerable concern in the village about the disruption which would come from the massive undertaking of removing the spoil heaps.

Discussions with the Coal Board were slowed by the 1984 miner's strike, when according to the Chair of the Heritage Committee 'everything stopped whilst the lawyers considered the strike' and 'when we eventually got around to buying it [for a nominal sum], they said well we will let you have it provided you take responsibility for the whole of the 5 acre site and not just the pit head'. The Local Authority recognized that the purchase of the site was a positive step but also a significant task for the Parish Council, and undertook to conduct the legal transactions with the proviso that the Parish Council took full responsibility for the site thereafter.

These negotiations were protracted and the site was finally transferred to the ownership of the Parish Council in 1987. In the meantime in efforts to secure the site and to make plans for its future, its status, which by then was in effect a mature woodland, was critical. The trees had not only created a particular place, they became central to efforts to secure it. Tree Preservation Orders were taken out on trees on both batches. In 1985, ACCES (the then County Council's community works programme) drew up a comprehensive survey and proposed management plan for the batch. This report pointed out 'that the site was a particularly interesting woodland as an example of what can develop on a sterile coal tip, given time and lack of harmful interference. Part of its function, indeed, could be a demonstration of such reclamation' (p. 3). The report stated that local knowledge placed the planting of the conifers in the 1890s, and recorded the presence of various self-seeded trees and shrubs. These included 'beech saplings dotted under the coniferous canopy and one dense thicket of ash saplings and on the north slope, and examples of English oak, turkey oak, holm oak, wild cherry, silver birch English elm, wych elm, Norway maple, sycamore, yew, rowan, holly, hawthorn, blackthorn, pussy-willow, elder, hazel, privet, gelder-rose and wayfaring trees' (p. 2). Such variety added weight to the report's recommendation that the site 'be maintained and enhanced as a developing woodland, and to create certain features (paths, glades, etc.) which encourage the public to appreciate it as such, but which do not detract from its wildness' (p. 4).

The pressure to see the site as woodland and best used for nature conservation and public access told, and the Parish Council suddenly had on its hands a large site mostly covered with mature conifers and other trees which had self-seeded and grown where space was available. (Figure 10.3 shows the batches as they are today.) The Heritage Committee was formed and early decisions were to treat the whole site as an opportunity for public access, as a site of commemoration of the mining industry which had produced the batch and which was still a powerful symbol of local identity, and to enhance the site's ecological status as a woodland. The presence of the trees presented both opportunities and difficulties.

The opportunities reflected the possibility of attracting funding in the form of woodland grants, and grants from other sources supporting nature conservation and public access to open space. The problems were that the trees at present made for a 'fairly user unfriendly' site which was 'very dark' (interview with the then Local Authority Tree Officer). The trees had been planted in quite a dense pattern and had 'not been managed as a forest for some generations – if ever – which has meant the original planted trees have grown very closely, very tall, with restrictive girth' (ACCES, 1985: 3). The

Fig. 10.3. The Camerton batches, viewed from the south-west (Owain Jones).

trees were thought to be 'nearing 100 years of age and subsequently the end of their natural life span' (Camerton Heritage Committee Report, 1993: 1). In response to this position,

> a scheme devised by the County Council Woodland Officer was drawn up to replace the old conifer trees on the batch with broadleaf species over a ten year plus period under the funding from the Forestry Commission, with the idea that sales of timber from the old trees would help offset the cost of redeveloping the site into an Industrial Heritage and Public Open Space/Amenity Area ensuring that the area would be retained for the benefit of local residents.
>
> (Camerton Heritage Committee Report, 1993: 1)

This scheme had derived some of its information and rationale from the earlier ACCES report, and was got underway with a local forestry contracting company being brought in who agreed to carry out work taking felled timber as payment.

'Little Switzerland'

Conifers, particularly as deployed in now notorious afforestation policies of the Forestry Commission, have generally had an extremely bad press, as exemplified by Massingham's (1988 [1951]) analysis of 'The Curse of the Conifers'. As Wright (1992) put it,

> We British [ ] have looked at those coniferous plantations and decided we do not like them. We have brewed up a frantic symbolism of revulsion around them. We deplore the dark world beneath the coniferous canopy [ ] those wretched fir trees are as deprived of individuality as people under communism.

The basis for such views is seen in part when Condry (1974) compares walking in an oak wood and in a coniferous forest.

> To be in an oakwood where all the birds are singing and the wildflowers are gay in the dappled sunlight of a spring day, is a beautiful experience. Then if you walk [ ] into an mature spruce plantation you get the full impact of what a gloomy, flowerless, silent and depressing place a conifer forest can be. (p. 132)

In other words, at the national level of cultural constructions of trees and woodlands, the conifer was seen in strongly negative terms, particularly when compared with 'native' broadleaf woodland. Partly in response to this The Forestry Commission by this time was giving particular funding emphasis to the (re)planting of broadleaf woodlands (Forestry Commission, 1985).

The scheme to clear-fell the conifers in stages and replace with broadleaf trees was then partly in response to the conditions of the site, but was also driven by the conditions of the Forestry Commission funding – 'we clear felled and replaced with broadleaf because that's The Forestry Commission's edict' (Chair of the Heritage Committee). These decisions were in turn set within the wider 'anti-coniferous' culture exemplified above.

However, once the felling of the conifers got underway there was an extreme adverse reaction within the village. One of the contractors said 'the moment we actually started work on the site we were accused of being the "massacrers" – the butchers'. The couple living in and running the then Post Office opposite the entrance to the site gave the contractors 'an ear full' every time they went into the shop, although they 'tried and tried and tried' [to explain the logic of felling and management plans]. This couple became so upset at what was happening that 'they sold up and moved away' (interviews with contractors).

Another public meeting was held, where the Local Authority Tree Officer was asked by the Heritage Committee to explain what was happening and he recalls that 'peoples' opinions were very strong, and their emotions, emotions were running very high [ ] their main objection seemed to be the fact that we were – their perception was that we were cutting down all the conifers and replacing with dominantly hardwoods'. The Chair added that 'it caused a tremendous amount of feeling [with] the Parish Council who were the saviours [having saved the batch from clearance] being the demons'. 'Everyone nearly lynched the Parish Council Chairman.'

The batches had become a local landmark and were nicknamed 'little Switzerland' because of the steep terrain covered in conifers. The objections to the clearing centred around the threat to this landscape identity; to the trees as a perceived habitat (to some); to the trees as 'representatives' of nature; and the trees a 'memorial' for the mining days. One of the most vociferous objectors told how it was the threatened loss of this loved landscape feature which motivated her.

> I know they may say that they are not native, but the conifers were very dramatic [ ] I used to have to struggle up there early in the morning to wait for the bus [to a point which has a view of the two batches] and sometimes it was misty and, you know those lovely old Japanese [ ] prints, and you see these little pointed things with the mist rolling around – [it was] quite like that sometimes in the mornings.

Another who objected said that she had done so on the grounds of nature conservation, saying,

> I was one of them what protested ... because there was so much wildlife along there and there's so many different sorts of plants along there and deer and buzzards, it's a haven for wildlife. You get so many different species of butterflies along there because the buddleia's took over the bottom.

The trees were also serving as markers of the industrial past to the older community and were thus still valued in terms of identity and memory. We were told the opposition had come mainly 'from the older generation'. The cutting of the conifers was a felling of memories, even though there was the promise of new trees on the site. As one of the contractors said 'I would call it almost

Fig. 10.4. A view of the old batch soon after felling on the middle section (A. Rankine).

like a memorial garden, because those conifers there represented the flowers on a grave of something that which is no longer [there]'. The links between the conifers and the past, and their on-going presence in the landscape meant that it was these particular trees which became the talisman of memory. These trees became the bearers of powerful local symbolic freights which bestowed on them a value that was not easily assessed by those looking from broader, more general, cultural, management or even ecological perspectives.

In general the felling of trees does have a strong (often negatively received) visual impact and this further added to the various reactions set out above. The site after felling and extraction of the timber looked 'an eyesore because it sort of looked like it had been ploughed through, it looked like a hurricane had hit it' (see Fig. 10.4). This jolt to those who valued the site as a tree-place comes partly from the temporal disjunctures between certain social constructions of nature, and the slower, longer time frames embedded in tree growth and consequently tree management processes.

The subsequent management of the site was clearly affected by this surprising outburst of feeling. The plans to clear-fell all the conifers were abandoned. This according to one of the contractors 'was a victory for the people of Camerton who have been resistant to taking down the trees' and this has put the Committee 'in a very difficult position'. The planned natural regeneration was augmented by additional planting of new broadleaf trees.

According to the Local Authority Tree Officer this was 'in many ways like a PR exercise because it made people see that we were actually replacing [the trees]'. The local school was involved in the new planting scheme as a means of making connections with the community. But on clearance of the conifers, the natural regeneration 'was phenomenal [,] trees just springing up everywhere' (contractor). Those responsible for the management of the site now feel they would have preferred to let this process replace the trees which were cleared, rather than the planting that was undertaken. As Harmer and Kerr (1995) point out, the chosen method of creating, or in this case re-creating woodland, particularly the choice of whether to plant or encourage 'natural regeneration', is a critical decision in terms of a number of factors. Not least of these is the offsetting of the higher costing of planting against 'the earlier visual impact it can achieve' (p. 125). Another issue is the composition of the new planting. The Local Authority Tree Officer said 'I didn't want it at the end of the day to look like an arboretum', but rather as 'natural as possible in the landscape' but in retrospect he felt the planted trees which included oak, ash and beech, did not quite fulfil this aim.

This pressure on the management of the site by local opinion about tree felling in general and the loss of the conifers in particular has led to a certain frustration in the professionals associated with its management. One saying 'The site is kind of a hybrid monster as far as I can – that's my own feeling about it, the hybrid monster which people seem happy to visit and [there is a] let sleeping dogs lie attitude about it now', adding, 'Camerton is more of a slog with the added hassle of the local opposition'. The Local Authority Tree Officer who initiated the management said 'I would dearly love it, but have never pulled it off [,] people's reaction has always been horror, during National Tree Week you have a whole area of trees cut down'. The point being that he felt that public opinion finds it very hard to see tree felling as anything but as a negative, destructive act, when in fact the history of woodland and forestry in general is about long-term cycles of growth, clearing or felling and regeneration (Mabey, 1980).

The site now has a circular path built on it. It starts from the old pithead where there is a large statue of a miner and an information point, then follows along the top of the tip to the far end where it descends the tip face by means of a long set of steep twisting and turning steps. The path then runs through a flat area at the foot of the tipping, now being managed as a picnic/open-space area by clearing some of the trees and undergrowth. The path then climbs up another flight of steep steps to re-join itself on the top of the batch and make a route back to the start. A series of information points are marked around the walk, and these correspond to a small booklet, available on the site, which highlights with text and drawings some of the historical and ecological features of the site. In October 1997 the site was designated as a Nature Reserve under the 1949 National Park and Access to the Countryside Act, and this adds yet another component to the site's political and cultural complexity. One issue is that one of the main human uses of the site, that of local residents walking their dogs, is reconfigured. Dogs now should be kept on leads. Humans (and their pets) are in fact part of the ecology of the site just as much as the trees (and wider nature) are part of the culture of the site.

Other funding and support has been attracted to facilitate the development of the site, ranging from input from the British Trust for Conservation Volunteers, Rural Action Trust, and grants from the Local Authority, but there remains some uncertainty and unease about the future of the site in some respects. In the professionals engaged in the planning and practise of site management, this lies both in the difficulties of squaring ecological and arboreal management requirements with the expectations and feelings of the local community, and in the fate of the site once the present, extremely dedicated and effective members of the Heritage Committee retire from their positions. Some members of the community also feel an uncertainty, which is reflected in the poignant observation made by one of the objectors to the initial felling, 'Any time I hear any sawing now I wonder what's going on'. Within this complex situation an ongoing stream of grants and other support have to be found to support development and maintenance costs.

Conclusion

In this chapter we have shown how this one area of woodland, this one 'place of trees' has 'evolved' through a complicated and contingent coming together of differing ecological, technological and cultural elements over time. These elements have changed and faded in and out of the picture as they combine and recombine over time. The Camerton site, as other places of trees, is a socio-ecological process, which is recognized through its distinctiveness as a place. Human actors tend to want to fix place, to form attachments. That is why most forms of idyll, according to Eisenberg (1998: 143) are ways 'of denying or declawing change'. The reasons for this must in part lie in the observation that

> familiarity and predictability are important for many people. There is a common desire to live in a place which is stable and orderly, where social interaction entails what George Herbert Mead called 'a conversation of gestures', gestures which are mutually understood.
>
> (Sibley, 1999: 115)

This presents the nature of places as a somewhat paradoxical and double-edged phenomenon, because while people try to fix them in their identity, places as Thrift (1999) states 'must be seen as dynamic, as taking place only in their passing'. This leads to an uneasy situation for those managing places of trees, because they are in part the result of unruly dynamic forces of the developing trees' life patterns, and the cultural associations grafted onto them. Such tensions can go relatively unnoticed, but when the forces and tensions build up enough to cause a rupture, like a shift along a fault line causing an earthquake, disruption and disquiet may occur.

In the case of the place we have briefly sketched out above, the trees of the site, those planted on the batch all those years ago, the trees which seeded themselves, and those now subsequently planted, are critical in the formation of, and in this case disputes over, place identity and as a consequence its management. Trees have material (size, form, longevity) and symbolic qualities which mean they are likely to be powerful players in the formation of place identity, and this brings, as our opening quote from Mabey suggests, a significant baggage to the management of tree-places.

In the case we have outlined the trees have been in multiple inter-relating roles. They have: been symbols of a past (industrial) heritage and bearers of community remembering; played out a complex role as being constructed as native or 'alien'; they have been a landscape feature and icon of local distinctiveness; and a site of nature refuge. Within this the trees themselves have not been the passive recipients of contesting social constructions. Their seeding and growing abilities, their materiality, their longevity have played an active role on the complex constructions and reconstructions of this particular place.

Acknowledgements

Thanks are due to those concerned with the management and maintenance of the Camerton Heritage Trail, particularly Mr T. Webber, Mr K. Taylor, Mr A. Rankine, and Broadleaf Forestry Contractors; residents of Camerton who agreed to be interviewed; Steve Preddy and the Cam Valley Wildlife Group; Dr Chilcott; and Mr J. Cornwell. We thank the editors for helpful comments on earlier drafts.

Notes

[1] 'Arbori-culture: the place of trees and trees in their place' is funded by the ESRC (Research Grant, R000237083), we gratefully acknowledge this financial support.

[2] Skinner's Journal 1803 – 1834 is the famous account of his time as rector of Camerton where he was often in open warfare with the mining community and their lawless 'godless' habits.

References

ACCES (1985) Camerton Old Colliery: management scheme for wooded batch and pithead area. ACCES, Bristol.

Arnot, C. (1999) Where There's Muck ... *Guardian,* Society Section (28 April).

Brown, P. (1997) Greening of the Black Country. *Guardian,* Society Section (3 September).

Buchanan, A. and Cossons, N. (1969) *Industrial Archaeology of The Bristol Region.* David & Charles, Newton Abbot.

Burgess, J. (1993) Not out of the woods yet. *Landscape Research Extra*, no. 12, Spring.

Callon, M. and Law, J. (1995) Agency and the Hybrid Collectif. *The South Atlantic Quarterly* 94, 481–507.

Cam Valley Wildlife Group (1998) Letter to Bath and North East Somerset County Council Planning Committee, in reference to applications 98/03185/OUT and 98/03234/REM.

Camerton Heritage Committee (1993) Camerton Heritage Project, Camerton Heritage Committee, Camerton (Nr. Bath, UK).

Casey, E.S. (1998) *The Fate of Place: A Philosophical History*. University of California Press, London.

Cloke, P., Milbourne, P. and Thomas, C. (1996) The English National Forest: local reactions to plans for renegotiated nature-society relations in the countryside. *Transactions, Institute of British Geographers* 21, 552–571.

Cloke, P. and Jones, O. (1999) Non-human agency – the purposive and relational actions of trees. Draft of paper available from authors.

Cloke, P., Milbourne, P. and Widdowfield, R. (1999) Partnerships and policy networks in rural local governance: homelessness in Taunton. *Public Administration* (forthcoming).

Condry, W. (1974) *Woodlands*. Williams Collins and Sons, London.

Down, C.G. and Warrington, A.J. (1971) *The History of the Somerset Coalfield*. David & Charles, Newton Abbott.

Eisenberg, E. (1998) *The Ecology of Eden: Humans, Nature and Human Nature*. Picador, London.

Forestry Commission (1985) *Broadleaved Woodland Grant Scheme* (revised November 1985), The Forestry Commission, Edinburgh.

Goodwin, M. (1998) The governance of rural areas: some emerging research issues and agendas. *Journal of Rural Studies* 14, 5–12.

Harmer, R. and Kerr, G. (1995) Creating woodland: to plant trees or not? In: Ferris-Kaan, R. (ed.) *The Ecology of Woodland Creation*. John Wiley, Chichester.

Harvey, D. (1996) *Justice, Nature, and The Geography of Difference*. Blackwell, Oxford.

Ingold, T. (1993) The temporality of landscape. *World Archaeology* 25, 152–174.

Latour, B. (1993) *We have Never Been Modern*. Harvester/Wheatsheaf, Hemel Hempstead.

Little, B. (1969) *Portrait of Somerset*. Robert Hale, London.

Mabey, R. (1980) *The Common Ground: A Place for Nature in Britain's Future?* Hutchinson, London.

Macmillen, N. (1990) *Coal from Camerton*. Avon Industrial Buildings Trust, Bath.

Macnaghten, P. and Urry, J. (1998) *Contested Natures*. Sage, London.

Massingham, H. (1988) *A Mirror of England: An Anthology of the Writings of H.J. Massingham, edited by E. Abelson*. Green Books, Bideford.

Massey, D. and Jess, P. (eds.) (1995) *A Place in the World?* Oxford University Press, Oxford.

Mele, R.A. (1997) *The Philosophy of Action*. Oxford University Press, Oxford.

Moffat, A.J. and Buckley, G.P. (1995) Soils and restoration ecology. In: Ferris-Kaan, R. (ed.) *The Ecology of Woodland Creation*. John Wiley, Chichester.

Murdoch, J. and Abram, S. (1998) Defining the limits of community governance. *Journal of Rural Studies* 14, 41–50.

NCB (1970) *Review of Research on Properties of Spoil Tip Materials.* Wimpey Laboratories Ltd, Hayes, Middlesex.
Peck, J. and Tickell, A. (1994) Too many partnerships ... the future for regeneration partnerships. *Local Economy* 9, 251–265.
Plumwood, V. (1993) *Feminism and the Mastery of Nature.* Routledge, London.
Rhodes, R.A.W. (1997) *Understanding Governance: Policy Networks, Governance, Reflexivity and Accountability.* Open University Press, Buckingham.
Rival, L. (ed.) (1998) *The Social Life of Trees: Anthropological Perspectives on Tree Symbolism.* Berg, Oxford.
Schama, S. (1995) *Landscape and Memory.* Harper Collins, London.
Sibley, D. (1999) Creating geographies of difference. In: Massey, D., Allen, J. and Sarre, P. (eds) *Human Geography Today.* Polity Press, Cambridge.
Skinner, J. (1971) *Journal of a Somerset Rector, 1803–34.* Oxford University Press, Oxford.
Thrift, N. (1999) Steps to an ecology of place. In Massey, D., Allen, J. and Sarre, P. (eds) *Human Geography Today.* Polity Press, Cambridge.
Watkins, C. (ed.) (1998a) *European Woods and Forests: Studies in Cultural History.* CAB International, Wallingford, Oxon.
Watkins, C. (1998b) 'A solemn and gloomy umbrage': changing interpretations of the ancient oaks of Sherwood Forest. In: Watkins, C. (ed.) (1998a) *European Woods and Forests: Studies in Cultural History.* CAB International, Wallingford, Oxon.
Williams, W.J. (1976) *Coal Mining in Bishop Sutton North Somerset, c. 1799–1929.*
Whatmore, S. (1999) Hybrid geographies: rethinking the 'human' in human geography. In: Massey, D., Allen, J. and Sarre, P. (eds) *Human Geography Today.* Polity Press, Cambridge.
Whatmore, S. and Boucher, S. (1992) Bargaining with nature: the discourse and practice of 'environmental planning gain'. *Transactions, Institute of British Geographers* 18, 166–178.
Wright, P. (1992) The disenchanted forest. *Guardian Weekend* (7 November).

11

Forest Tourism: Putting Policy into Practice in the Forestry Commission

Richard Broadhurst and Paddy Harrop

Introduction

This chapter looks at the background to forest tourism in the publicly owned woods of Britain, nearly all of them open to public access. It draws on the long history of open access to consider the approach of the Forestry Commission in planning and managing forest tourism, in promoting best practice across the forestry industry, and the development of policy.

Background

Values

Tourism is about much more than just forestry. Likewise, forestry is about much more than just tourism. Forest tourism is, nevertheless, of immense value. Indeed the UK Day Visits Survey (SCPR, 1998) commissioned by a consortium of the Countryside Recreation Network member organizations, suggests that an estimated 350m visits are made each year to woodland. With an average spend on each visit of £3.20, the estimated expenditure on woodland recreation amounts to more than £1bn. This only refers to day visits and so must barely scratch the surface. If you look at all the tourism literature promoting the countryside of Britain, many of the images included contain trees, woods and forests. What is the value of all these images, of all these trees: a proportion of all tourism receipts?

Work conducted (Benson and Willis, 1992) on the non-market recreational value of the Forestry Commission's forests estimated that these alone

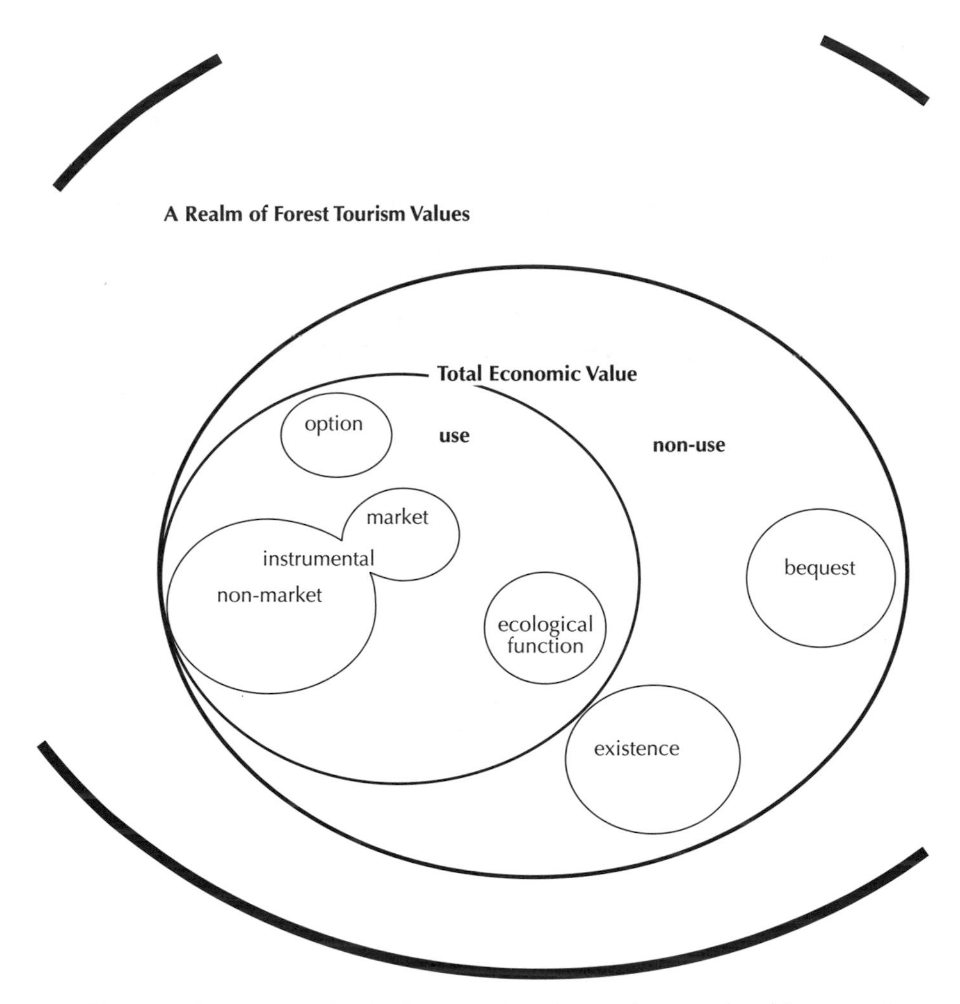

Fig. 11.1. Focusing on the Total Economic Value, within a realm of forest tourism values.

would account for £100m each year (1988 values). To derive the value, forests across Great Britain were clustered together, and samples chosen for investigation. The studies revealed a very wide range of values, which produced a mean value of £2 a visit.

The total value was computed crudely by multiplying the mean value of a recreational visit as derived through their work, by the estimated number of visits each year. Our experience suggests that the total forest tourism value includes much more than this. This (non-market) value needs to be added to the net income from the Forestry Commission's recreation activities which are priced. The resulting total is still just scratching at the surface of the contribution of publicly owned Forestry Commission forests to tourism, which extends beyond recreation.

Figure 11.1 enables the full range of economic values of forest tourism to be explored. The chart is one particular interpretation of work presented at a workshop on Environmental Economics, Sustainable Management and the Countryside, arranged by the Countryside Recreation Network (Bryan and Bateman, 1994). In the view presented here, and given a twist, the arrangement suggests that market values and non-market values are rather more closely related than is sometimes thought. There are those who seek to monetize non-market values or capture externalities, and others who seek to ensure that the countryside (in particular) and other marketable benefits are kept free, as they see them as public goods. Outside the sphere which represents economic values, the figure reveals that there is a realm of values. There is (most certainly) more than just economics. We know little about all the different components and how they add up (or otherwise interact) to create the total economic value, let alone the total forest tourism value. As a culture, we have been so clever at unpicking values generally that we often tend to underestimate large areas of the chart, and indeed frameworks other than economics. These may be just as important, if not more so, to society in determining what forests we should have and how we should manage them. For some of these frameworks, quantification is difficult if not impossible. Where it is possible, it may be inappropriate. Even so, there is a good deal of research to show that they are highly valued by society, and we should take care to describe the qualities so that we can take them fully into account in our decision making. These may be in many different guises, as the work of Common Ground in publishing a book of poems (King and Clifford, 1989) has shown.

History

To really understand the forests and forestry of Britain we have to think back, and retrieve our past. A history of our woodlands is a valuable starting point (Rackham, 1990; Smout, 1997) to set things in context and bring us to that crucial point in the development of forestry policy in Britain, 1919. We can only imagine what it must have been like at the end of World War I.

If you had been there, what would you have been talking about with your friends and loved ones? Forest tourism? We don't think so. You might have been dreaming of journeys you wanted to take, or having nightmares of ones you had undertaken, but everyone (everywhere) was more concerned with the basic essentials of life. During the war the necessary commodities were very hard to come by. Convoys across the Atlantic were under constant threat of torpedo attack. Supplies of timber were difficult to obtain. Wood and wood products are vital and required at all times. Just look around you, wherever you are, and imagine what would happen if you removed all wood and wood products from view.

Home-grown supplies were difficult to come by, too. By 1919, our woodland cover in Britain was down to just 4%. In these desperate times the

Forestry Commission was set up. The aim then was to develop a strategic reserve of timber through establishing national and private forests, just in case we should find ourselves similarly in trouble at some time in the future. Many countries have only really come to realize the value of their forested lands when, for whatever reason, they were about to lose them.

Fantastic efforts were made during the 1920s and 1930s to plant up barren hillsides, and areas which could be of no conceivable value to agriculture. Most of these new 20th century forests were planted in the uplands which, whilst of little agricultural value, had (and continue to have) enormous scenic and therefore tourism value. The greater the diversity and variety in such landscapes the more attractive the countryside.

The leading foresters of the day recognized that as these new forests were beginning to mature (at 15 years or so, newly established forests escape the worst risk of fire), they could be opened to enable visitors to enjoy recreation out of doors in the most scenic of areas.

The first Forest Park and the development of forest tourism

The first National Forest Park was established in Argyll, Scotland, in 1935, and others soon followed in areas now known as Glenmore, Queen Elizabeth, Kielder, Galloway, Coed y Brenin, Gwydyr, Dean, North Riding, and more recently Tay, Grizedale, Delamere, Thetford, Sherwood Pines, Whinlatter and Afan. The word 'National' was dropped later on to avoid confusion with the emerging National Parks of England and Wales. This was historic indeed. The first Forest Park was established well before national park legislation in England and Wales (1949) and in Scotland (expected 2001 or so). This was the beginning of the realization that these forest assets were more than just reserves of timber. These forests were resources rich beyond compare, generating social, economic and environmental benefits for local people, visitors, the nation and ultimately the planet.

By the 1960s, people most certainly did begin to think about recreation and tourism, in a way they had not before. Personal mobility, through rapidly increasing car ownership, was opening up untold freedoms, to the extent of causing concern amongst academics and planners (Dower, 1965). Although the envisaged excesses never materialized, there is in some areas significant pressure. The development of sustainable forest recreation and tourism can more easily meet the needs of people for more recreational opportunities, and of those planners and conservationists who are concerned about the pressure being placed on our natural environment.

The age of square-sided plantations on the hillside is a thing of the past. The work which foresters carried out to establish new forests under sometimes appalling conditions has had great effect. We must remember to judge them by the objectives of the day. Judged against these they succeeded in handsome measure. Many of these new 20th century forests are now being restructured, to provide greater variety. New forests are being designed with a discipline which will ensure that from the beginning we recognize the full range of stakeholders, and a greater range of benefits. More attention is now applied to the whole forest, not just the trees but the spaces in between, the water, the open ground, the rock-faces, the mountains, the glades, the rides and paths, the wildlife and the people.

It is clear that forestry today is about many benefits. The depth and range of benefits is only limited by our imagination. If we expand the circle of people we involve in developing policy, in planning and managing forests, we will enlarge still further the benefits to society.

The Forestry Commission and Multi-benefit Forestry

The Forestry Commission is the Government's department of forestry, working for all the people of Great Britain. It manages national assets (the publicly owned woods and forests) and also promotes best practice across the (forestry) industry as a whole.

The Forestry Commission's mission statement is to:

- protect and expand Britain's forests and woodlands and increase their value to society and to the environment.

The objectives are to:

- protect Britain's forests and woodlands;
- expand Britain's forest area;
- enhance the economic value of our forest resources;
- conserve and improve the biodiversity, landscape and cultural heritage of our forests and woodlands;
- develop opportunities for woodland recreation; and
- increase public understanding and community participation in forestry.

Although the Forestry Commission operates throughout Great Britain, its work is devolved with the exception of a few areas such as plant health, international policy, research and training. In Scotland, the Forestry Commission is answerable to the Scottish Executive and Scottish Parliament, and in Wales to the Welsh National Assembly. Joined-up government will be delivering policies to deliver forest tourism, with departments, agencies, the private and voluntary sectors all playing their part.

For almost 80 years the Forestry Commission has been welcoming visitors to the forest, sharpening its skills in managing the forest and in managing recreation and tourism. Because it has an unbroken chain of interest, the Forestry Commission is very well placed to research any forestry issue, to develop different approaches in the field, to refine them after feedback from managers, and then to promote ideas that work throughout the industry. Looking at each of these stages in turn:

Research

The Forestry Commission Research Agency has a major research station at Alice Holt Lodge, in Surrey, and another at Northern Research Station, Edinburgh, along with various out-stations and experimental (and demonstration) woods or plots. Much of the research is focused on forest science, and is concentrated on the life sciences. A growing area is the work on social forestry, where people and forestry interact, and the extent to which foresters can generate more social benefits. The Forestry Commission has also undertaken and collaborated in socio-economic work, to further our understanding of how people interact with forests.

Practice

With 10,000 square kilometres of forest, distributed in thousands of individual woods, there is scope for trying out different techniques, under different situations to see what works in practice, and also what does not. Forest Enterprise, the Forestry Commission's agency charged with managing the public woodlands, has something like 2400 staff, made up of foresters, civil engineers, land agents and administrative staff, who work as a large team to make things happen. With such a land bank, there is scope to try out approaches at different scales, across Britain or in a very localized way.

Refining practice

The Forestry Commission's broad scope allows it to test different management practices across the country, so best practice can be identified for different circumstances.

By the exchange of views about the efficacy of different management techniques, interventions can be refined through iterations with different teams in different parts of Britain, before wider adoption and field testing. As well as being the country's largest provider of countryside recreation, the Forestry Commission also operates a significant commercial recreation business. Its Forest Holidays division offers holidays in 166 forest cabins across four sites; 25 caravan and campsites spread across Britain from Deer Park to Strathyre, and the New Forest to Glenmore.

Promoting good practice

The Forestry Commission's Policy and Practice Division has advisers who can bring to bear the latest techniques used elsewhere. Their contact with foresters and managers throughout the world, as well as their practical experience of what works in Britain, ensures that their advice is up to date. The team includes a Principal Adviser on Social Benefits and Cultural Heritage, whose remit takes in recreation and tourism.

Planning and Managing Forest Tourism

Forest Enterprise manages recreation and tourism in harmony with other management objectives (Forestry Commission, 1998b) including producing timber for industry and preserving and enhancing the environment.

Managing forests in a sustainable way requires an integrated planning system to deliver a range of benefits to Government and the people of Britain. In this section, we describe how we implement strategies for managing forest tourism and give a case study of a well-visited sensitive forest in the Highlands of Scotland.

Open access

Forestry Commission land has a long history of open access. A key aim of our access policy is to extend free access on foot through as many forests and woods as possible. Cyclists are welcomed to many of the Forestry Commission woodlands distributed throughout Britain. As well as providing waymarked cycle trails we encourage the use of our extensive forest road network by cyclists. In 1999 we introduced a new policy for free access for horse riders to use forest roads and tracks in over 60% of Forestry Commission woods.

While providing free access for informal recreation, we also encourage organized sports such as: orienteering, rallies, husky dog racing and carriage driving through a system of permissions for events and competitions. These events and activities help to support local tourism initiatives, especially important in rural communities. The way we manage our forests also enhances the visual amenity and helps to attract visitors to the countryside.

Table 11.1. Recreation services.

Visitor centres	21
Waymarked walks	700
Forest drives	11
Cycle trails	2600 km
Forest cabins	166
Rangers	150
Events/Guided walks (each year)	600
Camping/caravan sites	25

What do we manage?

Forest Enterprise manages over one million hectares of land across Britain attracting approximately 50 million day-visitors (SCPR, 1998).

Forest Enterprise also manages a wide range of recreation services, summarized in Table 11.1.

We manage Forestry Commission land through our forest design plan process (Bell, 1998). To be successful, the Forest Design Plans must be written in the knowledge of the wider context and goals for the area, and must be kept alive and reactive to changes round about. The process requires nine steps:

1. setting objectives;
2. survey of the forest;
3. analysis of the information: constraints, opportunities and landscape character analysis;
4. concept design;
5. sketch design: felling, coupe design, timing, different silvicultural systems, restocking, choice of species, etc.;
6. documentation;
7. approval;
8. implementation;
9. monitoring and revision.

At a number of steps, there is scope to consult others who can contribute to the goals, or who may be able to achieve their goals through contributing to the plan.

Our aim is to produce a plan for the whole forest to deliver many benefits in a sustainable way. These 'forest design plans' will typically chart the next 40 years of planned management in outline and concentrate on the first 5 years in detail. They will usually be put together by a multi-disciplinary team and cover:

- recreation
- conservation
- timber production
- visual amenity.

Underpinning our work is the UK Forestry Standard (Forestry Commission, 1998c). It is our intention to work towards Forestry Stewardship Council (FSC) certification through the UK Woodland Assurance Scheme.

Much of the Forestry Commission estate has special conservation value. This short list gives an indication of the scale and importance of protecting these sites.

Sites of special scientific interest	62,000 ha
Ancient and semi-natural woodland	150,000 ha
Scheduled ancient monuments	1000

Glen Affric, a case study

Glen Affric is one of the largest surviving native pinewoods in the UK. It is managed by Forest Enterprise. A remnant of the pine forests that colonized the Highlands of Scotland 8000–10,000 years ago after the last Ice Age, it is home to a wide range of rare and endangered plants and animals including; capercaillie, pine marten, adder, twin flower, crested tit, large red damselfly.

Over 60,000 visitors come to Glen Affric each year to enjoy the forest and the surrounding area. Many of these visitors come from mainland Europe, and some from the USA and Japan. Tourism plays a major role in the economy of the Highlands of Scotland. The landscape and heritage of the area are among the main reasons for people choosing the Highlands as a destination. In 1994, Highland Council, Highlands and Islands Enterprise, Scottish Natural Heritage and the Forestry Commission formed a partnership. This was funded through the EU 'LIFE' initiative to develop the sustainable management of tourism in the Glen Affric area. Work has also been carried out in another partnership, with Trees for Life, Scottish Conservation Projects, Scottish Wildlife Trusts and Highland Birchwoods.

The objectives for the Glen Affric project are:

- to introduce visitor management techniques that will enable the reserve to accommodate current and future levels of tourist activity; and
- to enhance visitors' awareness and understanding of the environmental heritage in the reserve.

Table 11.2. Glen Affric, forest tourism management zones.

Zones	Description	Management policies
1. High intensity visitor management	Around the visitor centre development	Develop the visitor centre and orientation of visitors to main facilities Use international symbols and a range of languages
2. Medium intensity visitor management	Minor public roads and associated car parks and trails	Improve car parks and trails, and link interpretation to main reserve themes and visitor centre development
3. Low intensity visitor management	Forest roads and tracks	Improve car parks and trails, and provide few new facilities Inform visitors about the reserve and main facilities Emphasize mainly the natural environment interpretation linked to main reserve themes and visitor centre development
4. Core reserve with little, if any, intervention for tourism management	Quiet and undisturbed areas with no recreation facilities	Provide no new developments or information Allow visitors freedom to roam, but not encourage them

In order to focus the visitor management of Glen Affric and protect the sensitive environment, four zones were identified, in which visitors had different expectations and which required different management prescriptions. These were devised in order to ensure that the core reserve area would be protected and yet the forest tourism (and consequent flow of local economic benefits) encouraged to develop, and expand. Visitors are not prohibited from entering the core reserve area, but the provision of facilities in the other three zones acts as a magnet, and tends to keep visitors within the 'managed' area. The objective is to manage the recreation in such a way that visitors appreciate and enjoy freedoms, but in such a way that protects the core reserve area (Table 11.2).

A ranger service has been introduced to help visitors enjoy all aspects of the reserve. Informing visitors of the sensitivity of the site will help to protect it.

One role of the project is to monitor the impact of visitors. Survey programmes have been set up, but we recognize that this is a long-term project and changes should not be expected overnight.

Visitor numbers have been estimated using vehicle (and people) counters within the area. At present, the number of visitors appears to be stable. Surveys have also been carried out to monitor the profile of visitors to the area and to find out their perceptions and attitudes to what is intended and to what has been completed. Impact on the environment has been measured by collecting and weighing rubbish and by monitoring the long-term regeneration of the forest habitat. The final part of the monitoring work is concerned with looking at the length of stay and money spent by visitors in the area.

To deliver greatest benefit, tourism and forest management need to be integrated. Sensitive natural habitats are important in attracting visitors to an area, and in generating economic benefit. Forest habitats that support tourism must be protected to ensure the sustainable development of forest tourism. Managed sustainably, forest tourism can provide extra money for developing our woods and forests and at the same time increase their economic value to local communities. In this way, tourism and conservation become mutually supportive in the forest.

Promoting Good Practice

Information and technology transfer

Promoting best practice across the industry, the Forestry Commission is armed with an array of guidelines and practice guides which support the UK Forestry Standard. Although each guideline deals with a different area, for best effect, the actions have to be integrated. Training (Forestry Commission, 1999) provides the means to integrate such practice with existing skills and knowledge, and just as important is the less formal transfer of knowledge through networking.

To spread the benefits which arise from forest tourism, we need to do more to transfer the technology developed in forestry and in managing forest tourism. Arguably we also need to do more to research good practice. Much of the previous work of Forest Research has been concerned with forest science, but a significant body of work has been building up in recent years related to social forestry, and the field is developing.

Research

Research into the preferences and attitudes of people towards the forest for different activities revealed quite separate groups with different needs (Lee, 1990, 1999). Four major groups of visitors stood out:

- forest enthusiasts, who enjoyed the natural qualities of the forest, the fresh air, and the wildlife;
- day-trip makers, for whom the settings provided the opportunity for social interaction, appropriate places to take a family group or friends;
- sports enthusiasts, for whom the physical characteristics and settings matched the requirements of their chosen sport; and,
- dog walkers, for whom the forest provided the variety (and convenience?) needed.

Major attributes that attracted people included the presence of water, variety in setting, and diversity of species, colour, age of tree, the variety of life. People valued the relative solitude, peace and freedom, and the quality of naturalness associated with woodlands.

A study (Sime *et al.*, 1993) of the attitudes of owners in areas of England revealed a great diversity of views, from those who welcomed visitors (especially local people) to those who emphatically did not. The underlying central factor was the desire of owners for a degree of control over what activities people engaged in.

Looking at the provision for woodland recreation and access (Peter Scott Planning Services, 1997), it was found that the publicly owned Forestry Commission woodlands, which account for nearly 5% of Britain's land area, are almost all freely accessible. For other land, accessibility was more difficult to assess. In a survey conducted through the Timber Growers Association, in respect of 20% of other woodland, 58% of that woodland had 'at least some access'. Yet, in the accompanying case studies, it was suggested that in most woods in England and Wales access was deterred, or constrained to linear paths. The study suggested that what was required was a strategic approach for woodland recreation and access. We should reassess the manpower and specialist managers required, and also consider modifying grants. We should focus on further enhancing advisory services and developing the provision of information.

A study (Henwood and Pidgeon, 1998) in Wales revealed that people construe forests very differently, and that these different meanings held amongst different groups leads to communication failure. People also think and behave differently, as individuals and in groups. As individuals, we model our values not so much on economics but on deep personal meanings. When considering values to the community, though, we are more likely to consider economics. Other telling findings are that trees are seen as synonymous with nature, but that the forester is not necessarily seen as the custodian.

Another recent study (CSEC, 1998) identified the importance of local woods, close to where people live, in which provision is made for specific groups. Families with children need a quite different setting and suite of facilities and services than those without.

We need to remain mindful of different needs, identifying and meeting the needs of the different market segments. In all this, information is crucial (Future Foundation, 1998); ensuring that we know what people want, and ensuring that we provide information about the opportunities that exist.

What does the research suggest?

What does it mean for the way we manage forests? Quality is important. This is a message which comes through from many research and consultation exercises. The development of the UK Forestry Standard (Forestry Commission, 1998c), should ensure that all forestry grant-aided or managed by the Forestry Commission will meet the current criteria for sustainable management. There are some requirements in relation to recreation and access. Where recreation provision is made, it is expected that the Access standards agreed through the BT Countryside for All programme (Fieldfare Trust, 1997) should be applied, where appropriate. Beyond this the industry has been developing a voluntary assurance scheme for those who wish to demonstrate a chain of custody, to show that wood has been produced from forests which have been managed sustainably. The signing of an agreement to show intent to develop this as the UK Woodland Assurance Scheme has taken this process a stage further.

The delivery of quality service does not depend solely on assurance schemes. Working towards a shared understanding and towards shared values is essential to reach the highest levels of quality, of best fit between what the tourist wants and what the forester directly and indirectly supplies. Value is important too. Several pieces of work point to the value being computed in time as well as money.

Drivers for change

There are a number of strong currents and issues which are serving to shape the agenda of today. In the international context, it is notable that the

European Forestry Ministers, when they met in Lisbon in 1998, focused on the socio-economic aspects of forestry (Liaison Unit in Lisbon, 1998) and this followed extensive preparatory work (FAO/ECE/ILO, 1997). Within Britain there has been a great deal of work (Department of the Environment, Transport and Regions, 1998; Scottish Natural Heritage, 1998) looking at the implications of giving people greater freedoms (and responsibilities) in respect of access to the countryside, as new legislation is being considered. There is also a greater focus on supporting rural communities and a desire to make our transport more sustainable. It is within this context that policies and practices for sustainable forest tourism must develop.

Putting these ideas into practice, the Forestry Commission can develop the ideas pragmatically in managing the publicly owned forests, throughout Britain, and as exemplified by what is happening in Glen Affric.

It can also use other mechanisms. Roughly 40% of Britain's forests are publicly owned and managed by the Forestry Commission. The remaining 60% are owned by a wide range of organizations and individuals, some with a strong commercial interest, others where profit is a secondary concern.

Outside the Forestry Commission, grants (Forestry Commission, 1998d) and advice on good practice is increasingly important. The Forest Recreation Guidelines (Forestry Commission, 1992) still contain much useful information, and other guidelines in the series support the sustainable approach, which is drawn together in the UK Forestry Standard. With the development of the Forestry Commission's Internet (and Intranet) sites, we can ensure that people are kept aware of the latest research and practice, and remain up to date. Meetings and training events also play a crucial part in ensuring that ideas are exchanged (Forestry Commission, 1999).

Forest benefits are as diverse as the forests themselves. Amongst other things, forests provide a source of products and materials; a setting (within which to live, work or play); and an essential component of the ecosystem. Forest Tourism is an important benefit, because if we get it right, it will encourage more people to obtain more benefit from more forest, whilst giving more support to rural communities. To do this well, we have to start from visitors' perspectives. Visitors and forest users should play their part in developing forestry policy.

Developing Policy

Tourism policy

There are a number of Government Departments involved in developing tourism policy, and (of course) policy exists at every level from that of Europe, the United Kingdom, country and local authority. In developing the policy, the relevant level of government seeks the views of industry, commerce, finance, and of all corners of society. An example of such an approach is the recent consultation by the Department of Culture, Media and Sport into Sustainable Tourism. The Forestry sector plays only a minor role in this, as much tourism is dependent on our towns and cities, and the accommodation, attractions and transport sectors.

Forestry policy

As the department of forestry for Great Britain, the Forestry Commission plays the major part in advising on forestry policy, but seeks to be inclusive in developing such policy. One of the aims is surely concerned with how forestry can make the greatest contribution possible to tourism. There is a major role for the public forests which are widely distributed across our countryside, although rather more heavily concentrated in upland areas. There are a few notable exceptions of the New Forest, the Forest of Dean and Thetford Forest.

Woodland Grant Scheme

Within the Woodland Grant Scheme there is scope for grant aid towards the management of woodland for recreation (through Annual Management Grant) and for planting woodlands for community benefit (using the Community Woodland Supplement). Beyond that though the grants are concerned with the sustainable management of the woodlands we have and expanding our woodland cover. The benefits are released through management, to generate social, economic and environmental benefit.

Forestry strategies

The English Forestry Strategy, developed through extensive and iterative consultation (Forestry Commission, 1998a), includes Forestry for Recreation, Access and Tourism as a strategic priority and programme. Increasing access, improving the quality of information about access, enhancing the nation's forest estate, and promoting better understanding are identified as actions in the programme.

In Scotland, a working group of Government Departments and agencies produced 'Forests for Scotland: consultation towards a Scottish forestry strategy' (Forestry Commission, 1999). The document sought to stimulate debate and discussion about what people really want from forests. It seems that jobs, of all kinds related to forestry (including forest tourism), will feature highly in many of the submissions. The exact balance of social, economic, and environmental benefits sought from forestry in each area, and for each community, is likely to vary. The contribution which a relatively remote forest makes can take the form of the infrastructure for rural tourism. It might support a cycling hire company or provide activities to keep holidaymakers in an area for an extra day, and staying another night in a Bed and Breakfast. This may be every bit as important to that rural community as the much broader contribution of the Queen Elizabeth Forest Park to the proposed Loch Lomond and Trossachs National Park community. Jobs and parts of jobs are vital to support the social and economic fabric of rural (and indeed any) communities. In Wales, a similar consultation process is about to be launched.

During 1999 our strategy for recreation on Forestry Commission land will be developed. Consultation with other government agencies and departments, user groups, communities and FC staff will be the basis for the strategy. This will be combined with visitor survey and public perception of forestry information gathered over the last few years. The strategy will be based on needs of our current and future visitors and will help to link our work with other recreation providers.

Added to the energy being invested in national consultations is the realization of the role and importance of local communities, and the consequent need to ensure that consultation occurs at every appropriate level. The Forestry Commission is preparing its staff, through training courses, for engaging more closely with local communities.

Through greater communication we can generate even more benefits from forests, for forest tourism as well as all the other objectives people may have. There is a deliberate striving to consult communities more at local and national scale, through all manner of tools from Forest Panels, to Local Forestry Frameworks, to Indicative Forest Strategies at local authority level, to the national (not just Forestry Commission) strategies. This activity will provide conditions in which sustainable forest tourism can flourish. A deeper understanding of what people really want will underpin our success. It will be for all sectors to make the most of the opportunities which arise. Sharing information about best practice amongst foresters, tourism managers, students, researchers, planners, and all of the other communities of interest will be vital, to enhance the value of forest tourism. Through forest tourism we have the opportunity to join up our thinking and practice, so that tourism increases enjoyment as well as understanding of our relationships with trees, woods and forests, and the essential role they play in sustaining life on our planet.

References

Bell, S. (1998) *Forest Design Planning.* Forestry Commission, Edinburgh.
Benson, J.F. and Willis, K.G. (1992) Valuing informal recreation on the Forestry Commission Estate. Forestry Commission Bulletin 104, HMSO, London.
Bryan, F. and Bateman, I. (1994) Recent advances in monetary evaluation of environmental preferences. In: Wood, R. (ed.) *Environmental Economics, Sustainable Management, and the Countryside.* Countryside Recreation Network, Cardiff, pp. 3–23.
The Centre for the Study of Environmental Change (CSEC) (1998) Lancaster University, Woodland Sensibilities. A report for the Forestry Commission, Edinburgh.
Department of the Environment, Transport and Regions; and Welsh Office (1998) Access to the Open Countryside in England and Wales: a consultation paper. Department of the Environment, Transport and Regions, London.
Dower, M. (1965) Fourth Wave: the Challenge of Leisure, A Civic Trust Survey reprinted from the Architect's Journal, Civic Trust, London.
FAO/ECE/ILO Team of Specialists on Social Aspects of Sustainable Forest Management (1997) *People, Forests and Sustainability.* ILO, Geneva.
Fieldfare Trust (1997) BT Countryside for All Standards and Guidelines; A Good Practice Guide to Disabled People's Access to the Countryside.
Forestry Commission (1992) Forest Recreation Guidelines. HMSO, London.
Forestry Commission (1998a) A New Focus for England's Woodlands. England Forestry Strategy.
Forestry Commission (1998b) Forest Enterprise Corporate Agenda 1998–2001. Forestry Commission, Edinburgh.
Forestry Commission (1998c) UK Forestry Standard, The Government's Approach to Sustainable Forestry. HMSO, London.
Forestry Commission (1998d) Woodland Grant scheme pack. Forestry Commission, Edinburgh.
Forestry Commission (1999) Forestry Training Opportunities. Forestry Commission, Edinburgh.
Future Foundation (1998) Forestry Futures. A report for the Forestry Commission, Edinburgh.
Henwood, K. and Pidgeon, N. (1998) The Place of Forestry in Modern Welsh Culture and Life, a draft report to the Forestry Commission, Edinburgh.
King, A. and Clifford, S. (1989) *Trees Be Company.* The Bristol Press, Bristol.
Lee, T.R. (1990) What kind of woodland and forests do people prefer? In: *People, Trees and Woods.* Countryside Recreation Research Advisory Group, Bristol, pp. 37–52.
Lee, T.R. (1999) Forestry Commission Technical Paper 18 (in press). Forestry Commission, Edinburgh.
Liaison Unit in Lisbon (ed.) (1998) General declaration and resolutions adopted. Third Ministerial Conference on the Protection of Forests in Europe, Ministry of Agriculture, Rural Development and Fisheries of Portugal, Lisbon.
Peter Scott Planning Services (1997) Review of Provision for Woodland Recreation and Access, a report for the Forestry Commission, Edinburgh.
Rackham, O. (1990) *Trees and Woodlands in the British Landscape,* revised edition. JM Dent, London.
Scottish Natural Heritage (1998) Access to the Countryside for Open-air Recreation.

Scottish Natural Heritage's Advice to Government, Scottish Natural Heritage, Perth.

Sime, J., Speller, G. and Dibben, C. (1993) Research into the Attitudes of Owners and Managers to People Visiting Woodland, a report for the Forestry Commission, Edinburgh.

Smout, T.C. (ed.) (1997) *Scottish Woodland History*. Scottish Cultural Press, Edinburgh.

Social and Community Planning Research (SCPR) (1998) UK Leisure Day Visits: summary of the 1996 Survey Findings. Countryside Commission *et al.*, Cheltenham.

12

Recreation, Forestry and Environmental Management: The Haliburton Forest and Wildlife Reserve, Ontario, Canada

L. Anders Sandberg and Christopher Midgley

Introduction

Forest recreation in the province of Ontario, Canada, is typically focused in parks and preserved areas. There is very little integration of such activities with other economic pursuits, such as forest harvesting. Where integration occurs, it is usually as a result of logging in designated park areas, and it is generally frowned upon by the public, abhorred by environmentalists and carefully shielded from the visiting public by forest managers. For the most part, however, forest harvesting occurs on company-leased provincial Crown forest lands. On such lands recreational activities are not promoted generally, but may occur incidentally, such as is the case with fishing, hunting and snowmobiling. This is illustrative of the typical North American pattern of spatially separated uses (Sandell, 1998).

In the following, we tell the story of private forest owner Peter Schleifenbaum who is trying to buck this trend. The owner of the 21,751 hectare Haliburton Forest and Wildlife Reserve (hereafter HF; Fig. 12.1) in Ontario is attempting to build an environmental management strategy that combines recreational developments with the rehabilitation of a degraded forest and the cultivation of trees for value-added manufacture. His strategy is built on the effective use of the rights that go with private ownership, and a combination of other political, economic, public relations and market measures. We shall here describe these strategies and the operations of the HF, then try to assess their advantages, drawbacks and lessons for the combined use of recreation and forestry more generally.

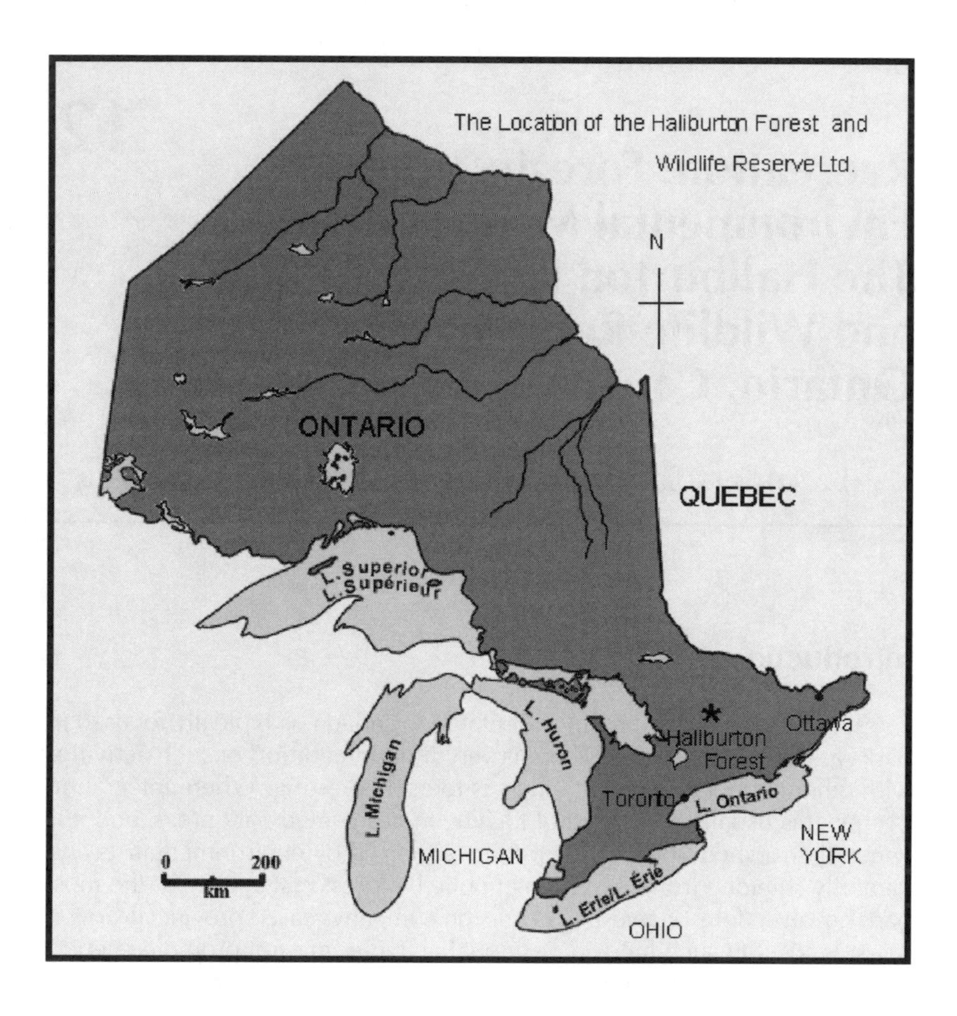

Fig. 12.1. Location map of the Haliburton Forest and Wildlife Reserve, Ontario, Canada.

The Setting

The HF is situated on the Precambrian Shield of central Ontario about 240 km north of Toronto, the largest Metropolitan area of Canada. The forest is part of the Great Lakes–St Lawrence Forest Region. Before forest exploitation began in the early 1800s, towering old-growth red and white pine dominated this forest region. These two species alone comprised 20–40% of the forest, with red and white oak, eastern white cedar, hemlock, black spruce, sugar maple, basswood, aspen, and white and yellow birch as lesser species (May, 1998; P. Schleifenbaum, Haliburton, 1999, personal communication).

Since forest exploitation started, large lumber and pulp and paper companies have high-graded the forest, taking the best trees and leaving the rest (Lower, 1973; Nelles, 1974; May, 1998). By the late 1950s, the forest structure had been changed substantially from only a century before. The forest was composed of less than 3% pine with largely successional sugar maple and beech in its place, much of which was of low commercial value (Stitt, 1994; P. Schleifenbaum, Haliburton, 1999, personal communication).

Faced with escalating harvesting costs, the last in a series of logging firms sought to sell the lands in the late 1950s. Essentially devoid of valuable timber and unsuitable for agriculture, the lands were considered practically worthless at the time, and after 4 years of looking for a Canadian buyer, the property was put up for sale in the European market. In 1962, a German, Baron von Furstenberg, bought the lands and gave the area its present name. After one year, Furstenberg sold the land to a fellow German, Adolf Schleifenbaum, whose family has since held on to the lands. A Canadian timber company, Weldwood, retained timber rights until 1970 when it ceased operations and shut down its mill (Stitt, 1994).

The Schleifenbaum purchase was not uncontested. During the late 1960s there was considerable debate as to whether or not such a large area of forest in Central Ontario should be privately owned. In fact, there was even a movement that sought to make the HF a part of neighbouring Algonquin Park – the 772,500 hectare provincial park that is the crown jewel in the Ontario Parks system – though it was short lived and had dissipated by the early 1970s.

Adolf Schleifenbaum's vision and management concept was informed by a European land-use model, which integrates rather than separates various forest-based activities. The first recreational activities developed were camping and snowmobiling. Camping sites were provided by abandoned lakeside log-landings, and snowmobiles accessed the network of logging trails (Haliburton Forest and Wildlife Reserve Ltd, 1977). Meanwhile, traditional economic activities were retained, such as logging, hunting and trapping, in spite of their growing unpopularity with North American recreationalists. From the very beginning, the vision of the HF has been to make such traditional activities compatible with, even supportive of, recreational activities. Indeed, an integral part of the HF strategy is to build understanding, and to provide education, for its vision. As Peter Schleifenbaum puts it, the HF's entire operation can be categorized as an outdoor education project.

The Recreational Facilities

Recreational activities are now closely and increasingly tied to other economic activities. Recreation accounts for 150,000 visitor-days in the HF and 64% of the total revenue. In the summer, camping is the most prominent and greatest revenue-generating activity. The backbone of the camping operation

Table 12.1. Visitor-days and operational breakdown of the Haliburton Forest (1998).

Activity	Visitor-days for recreation	Percentage of total revenue
Camping	40,000	22
Day Use	40,000	9
Wolf Centre	20,000	8
Snowmobiling	15,000	15
Hunting	5,000	3
Outdoor Education/Miscellaneous	30,000	10
Total Recreation Use	150,000	67
Timber	not applicable	18
Eco-Log	not applicable	12
Total Extractive Use	not applicable	30
Other	not applicable	3
Total	150,000	100

is a network of 320 semi-wilderness campsites that are well spaced and unserviced, yet accessible by car and boat via 17 of the HF's 50 lakes. Camping accounts for 40,000 visitor-days and 22% of total revenue (Table 12.1). Day users account for another 40,000 visitor days and 9% of the total revenue. They take part in a number of activities including mountain biking, hiking, canoeing, bird watching, orienteering and fishing.

One particularly popular attraction is a 6 ha wolf enclosure and interpretive centre that operates all year round. It provides educational lectures to groups of about 20 people every Saturday, an important facility in the Haliburton region given the long-standing local tradition of viewing wolves as vermin that should be eradicated. The centre is modelled after an American facility in Minnesota that hosts 60,000 visitors per year (Downs, 1997). So far, the HF's wolf centre hosts 20,000 visitors per year, and accounts for 8% of the total revenue.

There are other activities in the HF. These include a logging museum. There is also a restaurant and rustic accommodation available for up to 70 people at the base camp, the old site of the sawmill and logging operations. A canopy boardwalk, which gives visitors an opportunity to tour the treetops, is the most recent addition to the range of activities, and exemplifies yet another successful way in which education and recreation have been conjoined to draw people into the forest.

During the winter months, the HF hosts 15,000 snowmobilers, who have access to 300 km of what is regarded as some of the best trails in Ontario (P. Schleifenbaum, Haliburton, 1999, personal communication). The restaurant/bar and overnight accommodations at the base camp provide the focal point for the snowmobilers, but there are also huts throughout the forest to stop and warm up. Snowmobiling accounts for 15% of the total revenue.

The visitors to the HF come by car. Between 25 and 30% of the visitors come from the local area, within an hour's drive of the HF. The majority,

30–35%, come from the suburbs of Toronto, primarily those closest to the HF, about a 2-h drive. But there are also a significant number of visitors from Toronto (5–7%, 3 h away) and the more southern parts of Ontario (10–15%, more than 4 h away). This is likely because these areas are part of the Great Lakes Lowlands, which are largely cleared of trees, privately owned, and in agricultural or urban use. For residents in these areas, the HF constitutes one of the closest points for recreational activities in the lake and forest landscape of the Precambrian Shield. Fewer visitors come from northern locations, such as around Canada's capital city, Ottawa (4–6%, 3 h away), probably because of similar recreational facilities in closer proximity. Some visitors come from the United States, such as Ohio, New York and Michigan (3–7%), and 2% of the visitors to the HF are European.

Environmental Management

The environmental management of approximately 150,000 recreational visitor-days is potentially severe and has to be managed carefully. At the most general level, the HF is divided into three recreational areas: primary, secondary and tertiary. By guiding recreationalists to attractions and facilities, such as lookout points, picnic sites, observation and warm-up shelters, and self-guided walks, 95% of all visitors are funnelled into the base camp area, which constitutes only 2% of the area of the HF. The secondary recreational area constitutes 25% of the HF and receives 75% of all visitors. The tertiary recreational area, constituting close to 75% of the HF, receives less than 15% of all visitors. This leaves the human impact on the environment relatively concentrated and easier to monitor and manage. In the primary recreational zone, for example, the HF has developed a comprehensive household waste management system, and arrangements for the safe disposal of scrap metal and toxic wastes.

The environmental management of the HF as a whole consists of a careful inventory and monitoring of ecological change, and such management is often integrated closely with the hunting, fishing and forestry activities. Typically, wildlife management and wildlife studies are done in collaboration with other institutions. Both hunting and trapping are tied closely to wetland management through a combined effort with Ducks Unlimited (a popular North American Conservation organization that focuses on the protection of wetlands and waterfowl). The growth of poplar trees is promoted in certain areas as feed for beavers. The dams built by the beavers create wetlands for ducks. The ducks are hunted, the beavers are trapped for pelts, and their carcasses fed to the wolves at the Wolf Centre.

Students from Hocking College in Ohio return to the area annually to carry out research and present their findings on the ecology and wildlife of the forest. Trent University in nearby Peterborough has conducted a long-term study of the nature and spatial distribution of salamanders and ground beetles, data that are now part of a Geographic Information System. Students from the University of Toronto conduct studies in the HF; a doctoral study on biodiversity is soon to be finished. The Wildlife Service of Environment Canada also performs studies in the HF; among them are studies on the impact of global warming and acid rain, and the state of various endangered species. Additional studies have been conducted on the white-tailed deer; tracking their movement, and measuring the occurrence of brainworm in local populations.

There have been frequent explorations and experiments from experts in the fields of aquaculture and fish farming. At Lake MacDonald, the home of a genetically unique species of trout, a bass fishing tournament is actually a management plan designed to prevent excessive competition for the trout by the bass (an introduced species). The contest also provides an opportunity for ongoing research in to the fish stocks of the HF.

During the winter season, Schleifenbaum has taken several measures to mitigate the environmental impact of intensive snowmobile use. Damage to the ground caused by the compaction of snow is minimized by confining snowmobiles to prescribed trails, and these trails are (or once were) access roads for the logging operations (Neumann and Merriam, 1972). As such there is no understorey to destroy, and the ground is already far too packed for any burrowing animal. In fact, the use of all-terrain motorized vehicles when there is no snow cover has been banned in the forest for these very reasons. The lack of respect that was shown towards nature by users of such vehicles quickly led to their prohibition.

Through careful monitoring, Schleifenbaum has also found that fragmentation of wildlife habitat caused by the dense web of snowmobile trails may not be especially severe in the HF. Firstly, the species that characterize the region are moose, deer and wolves. These animals, in fact, utilize the trails as travel corridors, preferring to wander on packed snow rather than in deep snow. Furthermore, many of the roads that wind through the HF have been there for generations, thus any animals in the area would not know life without them, and the roads that do exist are kept as narrow as possible, with the tree crowns above them almost entirely closed.

While snowmobilers are not significantly bothered by the perpetual drone of engines (Badaracco, 1976), the impact of noise on the wild community is of concern. Some evidence suggests that the moose, deer and wolf population is oblivious to the noise. It may even be that the noise of human presence is comforting to the animals: that way the animals know how to avoid the humans. However, very little is known about the impact of noise on other fauna. Presently, the staff at the HF (including a wildlife biologist) are beginning to search the literature on this subject, and there is a suspicion that noise may be especially detrimental for the owl species found in the region, namely the Northern Barred Owl (*Strix varia varia*), the Northern Hawk Owl (*Surnia ulula caparoch*) and the Great Horned Owl (*Bubo virginianus virginianus*) (Johnsgard, 1988).

Another environmental impact of snowmobiling is its heavy reliance on fossil fuels, a factor that is exacerbated by the serious inefficiency of their two-stroke engines. The result is 'a large part of the fuel passing directly into the exhaust pipe and the environment' (Abrahamsson, 1998). Add to this the energy needed for (and the pollution generated in) the manufacturing, maintenance, and transportation of snowmobiles, and one finds that the level of pollution begins to grow in significance. In response to this point, however, it may be argued that the forest itself is a carbon pool and that the amount of carbon dioxide sequestered through the annual growth of the trees far exceeds the amounts produced by the snowmobiles (P. Schleifenbaum, Haliburton, 1999, personal communication).

An argument in support of localizing snowmobiles to a particular area such as the HF is that it diverts considerable snowmobile traffic from free access trails that are regulated less stringently. If the thousands of weekly visitors to the HF took up snowmobiling in the more pristine wilderness areas, there could be an increased incidence of environmental damage by people taking their machines off the trails and into the sensitive areas. Also, confining snowmobiling to special areas such as the HF which are well maintained and patrolled may decrease the incidents of human injury and death in a recreational activity that is notoriously dangerous (Canada Safety Council, 1971; Rabideau, 1974; Rowe *et al.*, 1992; Gray, 1998). Overall, snowmobiling may be a necessary evil for the HF. Its negative effects on the local fauna and flora will always be measured against the revenue accruing from snowmobiling.

Since 1989, when Peter Schleifenbaum replaced his father as manager and operator of the HF, the integration of environmental management with forestry, recreation and outdoor education has been a primary goal. Tree harvesting follows the low-grading principle, leaving the best and taking the rest. Schleifenbaum also uses a long-trusted field approach: once crown closure reaches 80–100%, an area is considered ready for harvest and individual trees are selected and cut, reducing crown closure to anywhere from 30 to 60%. Standard calculations suggest that the HF can yield 10 m^3 ha^{-1} $year^{-1}$, while only 2.5 m^3 ha^{-1} are harvested from the HF in a typical year (P. Schleifenbaum, Haliburton, 1999, personal communication).

Another significant forest operation in the HF stems from a 1995 windstorm that totally destroyed 600 ha and moderately damaged another 800–1200 ha of the forest, blowing down much of its eastern hemlock. These fallen trees constituted the impetus for an ongoing commercial venture, Eco-Log, which provides log building kits with full instructions for assembly. The venture is marketed as a concept combining the use of wood from sustainable forest management with economic considerations offering an alternative to common log-building concepts.

In extracting small amounts of low quality fibre or salvaging storm-felled trees, forestry in the HF is not profitable in the short term. However, Schleifenbaum estimates that in thirty years the trees harvested will yield a

90% sawlog component, up from the 30% that exists now. The significance of this is that sawlog trees are ten times more valuable than pulp trees on the market. The improvement in the quality of the forest promises to increase the revenues from forest products. Presently, forestry accounts for only 18% of the total revenues generated by the HF (Table 12.1).

Increasing the value of the forest, however, has a deeper meaning than simply elevating the commercial value of the timber. It also means attributing value to threatened species that exist in the HF by attempting to restore them. A good example centres on the severely depleted pine species. When a 10 ha red pine stand was discovered, it was set aside as a protected area from which no trees can be taken. There are also plans to experiment with small clear-cuts (up to about 1 ha each). These clearings will create the openings in the canopy necessary for planted seedlings, which require a great deal of light, to take hold. This re-introduction reveals an effort to restore the diversity that historic logging practices have diminished. Although it will take several decades, pine trees may once again become a significant biological component in the HF, which is illustrative of Schleifenbaum's commitment to rehabilitate the long-term health of the forest. In collaboration with the Forest Engineering Research Institute of Canada, the HF has pioneered the testing of more sustainable and safer harvesting methods. Recent tests have involved truck safety on logging roads and the use of lightweight ropes rather than heavy cables to pull logs longer distances from the cutting sites to the logging trail.

The promotion of sustainable forestry and hunting and trapping is part of the recreation and outdoor education strategy. The selective harvest regime employed at the HF is designed to maintain the structure of the forest. Thus, the forest remains aesthetically pleasing for visitors even while logging provides 30% of the revenue of the HF and employs 20 full-time contracted workers. Guided tours inform and educate about the cutting regimes and the rehabilitation efforts. At one point, Schleifenbaum even maintained a horse-logging operation that turned out to be more valuable as a tourist draw than as a logging operation.

Perhaps the most important step taken by the HF to promote itself as a multiple-use recreational facility relates to the issue of forest certification. One reason for this is that those who come to the forest know that they are supporting an operation that has been recognized as an important element in the health of both the local environment as well as the local economy. In 1998, the HF was the first forest operation in Canada to seek and gain certification for sustainable forestry operations under the umbrella of the Forest Stewardship Council (Cabarle *et al.*, 1995). The FSC is an international non-government organization that has developed a set of principles and criteria for sustainable forestry. It also accredits certifiers that provide third-party independent audits of the management of forest assets. Environmental criteria are clearly important. But equally significant are the criteria concerning community relations, workers' rights, and economic and social benefits.

Ultimately, those who seek certification hope to make their products more appealing to environmentally conscious buyers and consumers.

A non-profit certifier from New York State, Smartwood Inc., performed the forest audit of the HF. Smartwood characterized the area as a well-integrated, multiple-use operation that is very accessible to the local community and is a good source of certified wood (Smartwood, 1998). Some of the problems identified were the absence of a standard forest management plan and the lack of knowledge about the impact of snowmobiling on the local flora and fauna. The lack of an official management plan also prompted various attacks from rival certifiers who argued that without a plan it is impossible to ensure that the forestry is definitively sustainable (Armson, 1998). The certification nevertheless stuck with the proviso that a management plan be provided within a year.

While Schleifenbaum expects most of the benefits of forest certification to be realized in the long term, he has already seen some gains. The hardwood harvested in the HF is sold to a local lumber mill owned by Tembec, who manufactures hardwood flooring under the certification label of the FSC. The HF also now supplies pulpwood to the Lyons Falls Paper Company in upstate New York. This small paper manufacturer was the first company in the United States to produce writing, printing and specialty paper without the use of chlorine. They also use high-yield pulp which requires fewer trees per ton of paper. With this environmental record, the availability of wood coming from a forest accredited as well-managed provides additional support to the paper company's claim to environmental stewardship. Indeed, the Lyons Falls Paper Company has become the world's first producer of printing paper guaranteed to come from a sustainably managed forest, a condition that is expected to grow in importance as consumers begin to seek out products less damaging to the environment (National Wildlife Magazine, 1999). Meanwhile, the HF has acquired the bragging right to tell their visitors about their sustainable forest operations, and to use them as an enticement for potential recreationalists.

In spite of their growing importance, the forestry operations in the HF are still a small part of total revenue and employment (Table 12.1). Recreation is and no doubt will remain the dominant activity. Schleifenbaum has had to make this point in dramatic fashion in the face of government policy. From the early 1970s to 1994, private owners of forested lands exceeding 40 ha benefited from a Managed Forest Tax Rebate, which rebated up to 75% of property taxes for lands under forest management. In 1994, the provincial New Democratic Party cancelled the Managed Forest Tax Rebate, arguing that this was necessary in a time of growing government budget deficits. Woodlot owners, however, felt that the cancellation constituted a political cash grab for the government, with some woodlot owners facing a quadrupling of their tax bill. In response to this, they began to cut off access to the land for the public, threatening the booming recreation industry in the Haliburton area and the jobs that came with it (Walker, 1994).

Schleifenbaum in particular exerted considerable leverage by clear-cutting 15 ha of forest land adjacent to cottage properties. The cut generated 8000 dollars for the HF, but cost one owner, who was trying to sell his cottage, 40,000 dollars in property value. Thus 15 ha of harvested timber had a fifth of the economic value of the aesthetics of the living stand. While the official argument was that Schleifenbaum needed to start clear-cutting in order to make enough money to pay his taxes, he also illustrated quite effectively that trees have recreational value. Thus, somewhat ironically, it can be argued that this bold move was done in the interest of the forest (i.e. clear-cut 15 ha right away in the hopes that it never has to be done again; P. Schleifenbaum, Haliburton, 1999, personal communication). As a result of the subsequent public outcry, the Progressive Conservative Party pledged to reinstate the Managed Forest Tax Rebate. This occurred in 1996.

Haliburton Forest and Wildlife Reserve Assessed

In examining the nature of outdoor recreational developments, Klas Sandell identifies two distinct strategies: dominant and adaptive eco-strategies. The eco-strategy of domination seeks to adapt nature to recreation with the help of technology and commodity exchange. A metaphor to describe this strategy is a factory, where large amounts of labour and energy are employed to create recreational products and services (Sandell, 1998). Typical activities include downhill skiing and indoor skating. From the perspective of the human relationship to the environment, this trend promotes a functional specialization of people's relationship with nature (Sandell, 1998). The eco-strategy of adaptation, by contrast, represents an endeavour to adjust recreational activities to local physical and cultural environments. Here there are two subcategories. Active adaptation seeks to raise the productivity of local landscapes with the help of technology, but takes as a point of departure the local territory. Such activities may include cross-country skiing, hiking, and skating on lakes, activities which require minor alterations to the landscape. The strategy of passive recreation, alternatively, is sceptical of raising the productivity of landscapes and suggests human subordination to nature (Sandell, 1998). A metaphor to depict this strategy is the museum, where no manipulation of nature occurs. Such activities include bird watching and nature walks.

The HF contains a tenuous mixture of all these strategies. Some aspects of snowmobiling fall within the dominant eco-strategy. Many snowmobilers in the HF are not from the local area, but travel for long distances to reach it. For most snowmobilers, the HF is a playground that exists for enjoyment rather than a place of interaction with the natural world. Nature takes on an instrumental value, where riding at high speeds through the forest can represent a triumph of humanity over the natural world. Even the engine noise drowns out most other sounds of the forest. The enjoyment of riding the machine is central and nature itself is peripheral. Also, in order to maintain ideal conditions for snowmobiling, considerable grooming, tending and patrolling of the trails is necessary.

But the HF also contains eco-strategies of adaptation. In both forestry and recreation, Schleifenbaum has raised the productivity of the landscape, but without altering it significantly. His forestry activities seek to respect and restore the integrity of the local forest ecosystem, and many recreational activities are banned or adapted to the local natural and cultural environment. Even snowmobiling is confined to logging roads that have existed for over a century.

Finally, there are elements of the passive adaptation strategy in the HF. Although they tend to play a lesser role as financial injections in the HF, hiking, canoeing, bird watching and listening to the wolves also bring in many visitors. By their nature they are low intensity and seek to immerse the human into the greater natural setting. Typical of the passive strategy, nature untouched becomes the playground and the requirement of modifications is eliminated. The particular activity is specifically designed to exist within the confines of what nature has provided on its own.

The educational activities often represent passive strategies of recreation. Aside from the wolf centre and the canopy boardwalk which rely on the existence of considerable infrastructure, and arguably perpetuate the objectification of nature and its inhabitants, the lectures and presentations generally only involve the absorption of knowledge on the part of the participants. Rather than consumption of the natural world, the educational activities create, for the most part, reflective involvement with the natural world. This is characteristic of the activities enjoyed by amateur ecologists and field naturalists.

The uniqueness and promise of the HF is the presence of the adaptation strategy, the combining of non-intrusive forestry operations that also allow for extensive recreational activities into a comprehensive environmental management plan. This might even be further reinforced if snowmobiling is scaled down (because of its harmful environmental impacts), and the revenues from forestry and adaptive and passive recreational activities continue to grow.

The HF provides both similar and different trails with the broader Ontario recreation scene. The HF is different because it is an island of private ownership in a sea of provincial Crown lands. By owning the forest in fee simple, Schleifenbaum is free to charge his visitors any price, a practice that cannot be exercised as aggressively for public lands (though more so recently). At 15 dollars for a day pass, or 190 dollars for the season, snowmobiling carries the HF through the winter months. A tour on the canopy boardwalk costs 65 dollars per adult, and mountain biking or canoeing requires rental or an investment, the drive to the HF from the city, and a ten dollar day pass. For the future, the HF is planning to develop more upscale accommodations to capture the emerging niche market in wilderness retreats for business clients. More and more, recreation, environmental enjoyment and outdoor education come at a price. From one perspective, the charge on access to recreational activities may be seen as the necessary price to pay for forest sustainability. The benefit is that the cost of sustainability is borne by those individuals who choose to go to the area for recreational purposes rather than the public at large. On the other hand, it has implications for the overall public access to recreational activities. Clearly, those who cannot pay are not welcome in the HF.

The HF also reflects some common features of the overall outdoor recreational landscape of Ontario. On the one hand, snowmobiling is an extremely popular activity in the province generally as well as in the HF in particular. Given the economic growth it generates, and the spendthriftiness of snowmobilers, it is likely to continue to receive widespread support in spite of the various forms of pollution it generates. On the other hand, large parts of the Ontario rural landscape are being gradually gentrified. Though at the margin of the more popular Muskoka region, the Haliburton Highlands region has still seen much summer cottage development, the growth of summer homes and an increase in the demand for passive recreational activities from southern Ontario urbanites (Wadland and Gibson, 1998).

Both of these processes are at once competing and complementary. As a dominant eco-strategy, snowmobiling is often in conflict with adaptive eco-strategies such as cross-country skiing and snowshoeing. On the other hand, both activities respond to social groups that have the ability to pay. More and more lands are set aside, enclosed and access restricted. Public access to recreational lands, so championed in the past, is thus increasingly being constrained (Killan, 1993).

The broader context of the HF case study, then, supports the more general trend toward increased spatial divisions and privatization of use in the Ontario countryside. Though Ontario residents are split over its significance, a recent province-wide exercise entitled Lands for Life confirms a similar trend. It has established that 12% of all Crown lands be set aside as preserved areas (essentially being available for active and passive recreational activities). Another policy initiative has delegated that parks and preserves set their own entrance fees based on demand, thereby making them operate more like businesses than a public service. The rest of the Crown lands, meanwhile, are assigned to intensive resource use by forest and mining companies, or market-based recreational businesses.

Conclusions

In the early 1960s, the Haliburton Forest and Wildlife Reserve looked very similar to many other Canadian private forest properties. The area had been owned by a successive number of forest companies which had high-graded the area for the most valuable timber. Little commercially valuable timber was left and the property was bought at a cheap price.

Since the 1960s, however, the HF represents an exceptional case. Continuing his father's vision, Peter Schleifenbaum, as owner-operator and professional forester, has cast a wide net to rehabilitate his property. The infrastructure built to support forestry in the area over 150 years has allowed for considerable recreational access to the forest. In addition, a rehabilitated forestry operation is designed not only to elevate the quality of the forest from an economic standpoint, but also to enhance the natural aesthetics and integrity of the ecosystem. These factors allow the recreational and educational activities to be compatible with forestry in the HF. Now, recreational and forestry components are integrated rather than separated ingredients of the operations.

The advantages of a privately owned forest to the health of the forest ecosystem are several. The owner can make quick decisions and implement them with speed. The compromises which typically plague public decision-making processes, and which often compromise ecological integrity, are absent. Private ownership may also instil forest stewardship, something which has even been extended to some of the campers who lease lands in the forest.

The HF has also been able to integrate successfully sustainable forestry and recreational activities, though with a few wrinkles. First, the extensive use of snowmobiling may not only be potentially harmful to some flora and fauna, notably owls, but it is also dangerous, and the snowmobiles' two-stroke engines generate considerable pollution. The HF also represents the enclosure of what in the past has been regarded as a forest commons. Access to the forest and its various activities is provided on a fee-for-service basis, and those who cannot afford to come are not welcome.

In some ways, the HF represents a larger process within the Ontario outdoor recreation scene: the spatial separation and privatization of such services. The recent political decision to set aside 12% of provincial forest lands as park and preserved areas, and to charge higher and variable fees for park entry, while leaving the rest to industrial forest activities, is part of this process. While the HF addresses two important issues: the promotion of sustainable forestry and the provision of recreational services, it does not (and probably cannot) deal with the question of providing access to recreation and environmental awareness for an even wider public (Utting, 1993).

Acknowledgements

We would like to extend our thanks to the Faculty of Environmental Studies at York University for a research grant in support of this project. We are also grateful to Peter Schleifenbaum and the staff at the Haliburton Forest and Wildlife Reserve who gave generously of their time and knowledge.

References

Abrahamsson, K.V. (1998) Sounds of silence? The dispute over snowmobiling in the Swedish mountains. In: Sandberg, L.A. and Sörlin, S. (eds) *Sustainability – The Challenge: People, Power and the Environment.* Black Rose Books, Montreal, pp. 121–129.

Armson, K.A. (1998) Forest Certification. *The Forestry Chronicle* 74, 284.

Badaracco, R.J. (1976) ORV's: Often rough on visitors. *Parks and Recreation* September, 32–35, 68–74.

Cabarle, B., Hrubes, R., Elliot, C. and Synnott, T. (1995) Certification accreditation: the need for credible claims. *The Journal of Forestry* 93, 13–16.

Canada Safety Council (1971) *Snowmobile Accidents in Canada: Fatal Accidents and Fatalities. Winter of 1970–1971.* Canada Safety Council, Ottawa.

Downs, P. (1997) A view into the domain of wolves. *Wolves* 1, 14–15.

Gray, J. (1998) Death by snowmobile: winter carnage in the country. *The Globe and Mail,* 14 February, Toronto.

Haliburton Forest and Wildlife Reserve Ltd. (1977) Camping, Trout Fishing, Deer Moose and Bear Hunting on the Private Lands of the Haliburton Forest and Wildlife Reserve Ltd., 100 Square Miles of Forests and Lakes. Haliburton Forest and Wildlife Reserve Ltd, Haliburton (map).

Johnsgard, P. (1988) *North American Owls.* Smithsonian Institute Press, Washington, DC.

Killan, G. (1993) *Protected Places: A History of Ontario's Provincial Parks System.* Dundurn Press, Toronto.

Lower, A.R.M. (1973) *Great Britian's Woodyard: British America and the Timber Trade, 1763–1867.* McGill-Queen's University Press, Montreal.

May, E. (1998) A*t the Cutting Edge: The Crisis in Canada's Forests.* Key Porter Books Ltd, Toronto.

National Wildlife Magazine (1999) NFW members at work: NFW teams with Lyons Falls Paper Company to pioneer Smartwood paper production. *National Wildlife Magazine* 37, 62–63.

Nelles, H.V. (1974) *The Politics of Development: Forests, Mines, and Hydro-Electric Power in Ontario, 1849–1941.* MacMillan, Toronto.

Neumann, P.W. and Merriam, H.G. (1972) Ecological effects of snowmobiles. *The Canadian Field-Naturalist* 86, 207–212.

Rabideau, G.F. (1974) Human, machine, and environment: aspects of snowmobile design and utilization. *Human Factors* 16, 481–494.

Rowe, B., Milner, R. and Bota, G. (1992) Snowmobile related deaths in Ontario: a 5-year review. *Canadian Medical Association Journal* 146, 147–152.

Sandell, K. (1998) The public access dilemma: the specialization of landscape and the challenge of sustainability in outdoor recreation. In: Sandberg, L.A. and Sörlin, S. (eds) *Sustainability – The Challenge: People, Power and the Environment.* Black Rose Books, Montreal, pp. 121–129.

Smartwood (1998) Public certification summary report for natural forest assessment of Haliburton Forest and Wildlife Reserve Ltd. Certificate #SW-FM-033. Richmond, Vermont.

Stitt, M.P. (1994) *The Forest and the Trees: Historical Roots of the Haliburton Forest and Wildlife Reserve.* Haliburton Forest and Wildlife Reserve Ltd, Haliburton.

Utting, P. (1993) *Trees People and Power: Social Dimensions of Deforestation and Forest Protection in Central America.* Earthscan, London.

Wadland, J. and Gibson, A. (1998) Learning a bioregion: Trent University and the Haliburton Highlands. In: Sandberg, L.A. and Sörlin, S. (eds) *Sustainability – The Challenge: People, Power and the Environment.* Black Rose Books, Montréal, pp. 177–188.

Walker, W. (1994) Tree-cutting blamed on NDP tax move. *The Toronto Star,* 2 November, A 11.

13

Writing an Environmental Plan for the Community Forest of Mercia, England

Graham Hunt

Introduction

Increasingly, as environmental issues are becoming recognized as an integral part of managing change in our environment, there is a need clearly to state policy objectives and outline plans of action. Until the introduction in the UK in the late 1980s of the requirement to provide an Environmental Impact Analysis as part of major development proposals, the majority of environmental plans tended to be site specific or generally related to conservation issues. It was perhaps only with the acknowledgement by national governments of the need to secure sustainable development options that environmental issues have received the recognition and status that they deserve.

This paper will focus on the methodology used to produce an Environmental Plan for one of the major initiatives of this decade. It will use the plan produced for the Forest of Mercia, one of 12 Community Forests promoted by the Government, as an example of a fully integrated plan-making process that was specifically designed to allow environmental issues to be considered in their proper local and national context. The discussion will highlight the two main options for plan making. It will outline in detail the methodology used in the production of the Forest of Mercia Plan and describe the benefits that resulted from choosing that particular option. Reference will also be made to other potential applications of approach adopted for the Forest of Mercia Plan.

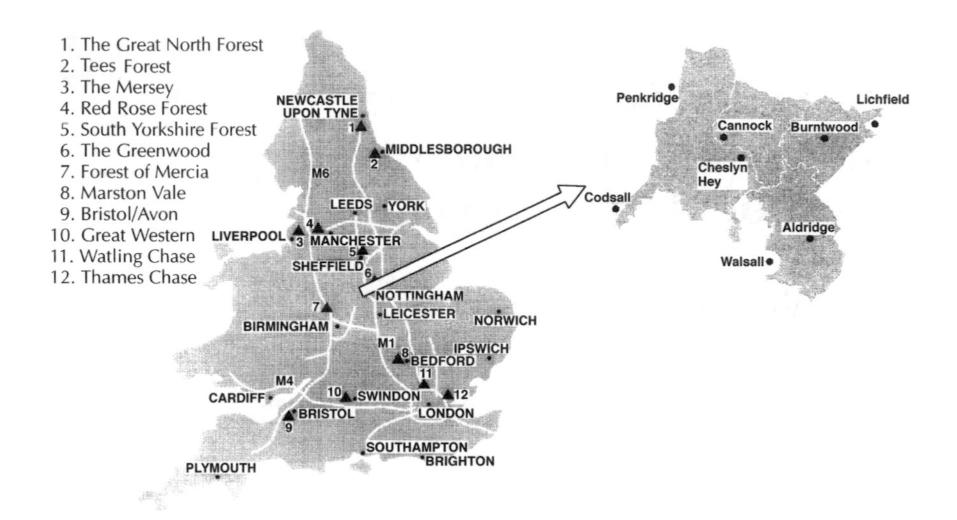

Fig. 13.1. Community Forests in England, and map of the Forest of Mercia area in the British Midlands. Sources: The Forest of Mercia Forest Plan (1993) and Annual Report (1998).

The Forest of Mercia

The Forest of Mercia Plan has been chosen as a case study for this paper because it is part of the largest environmental initiative in the UK at present. The UK Government had decided to embark upon a programme of new forest creation in response to a paper prepared by the Countryside Commission in 1987. The paper was entitled 'Forestry and the Countryside' and brought forward suggestions for the creation of new forests around our major cities as well as the creation of a new national forest. It is this policy document that provides the intellectual arguments for both the Community Forest Programme and the National Forest.

The Countryside Commission saw the use of forestry in the English lowland as an opportunity to bring about robust and sustainable landscape enhancement which would also bring a wide range of social and economic benefits to the communities that these new forests served. The principle of multi-purpose forestry was warmly endorsed by the Forestry Commission who on their own estates had for a number of years been successfully combining timber production with provision for recreation, conservation and environmental education.

The two Commissions decided to work in partnership and in 1990, with the full endorsement of the UK Government, embarked upon a pilot programme of three new forests. The Programme was entitled the 'Forests for the Community Programme' and contained proposals for the establishment of three Community Forests situated to the east of London, in the West Midlands and the north-east. The West Midlands Community Forest was quickly embraced by local people whose choice for the name of the project was the Forest of Mercia. Within 12 months of starting this pilot of three forests, the Government at that time decided to extend the programme to the current 12 forests (Fig. 13.1). These forests are primarily being targeted at the edge of major towns and cities in England and are generally designed to deal with long-term landscape problems caused by mining, extraction, urban expansion, industrial activity and rationalization of farming.

The Forests for the Community Programme was an integral part of the then Government's commitments made at the Rio Earth Summit. Their intention was to double tree cover within England. At the current level of 7% it is well below the European average. The incoming Labour Government have also warmly endorsed the concept of creating Community Forests and have committed themselves to increasing tree cover significantly within the United Kingdom.

In each designated forest area, the Countryside Commission and Forestry Commission have formed partnerships with the relevant local authorities. A Memorandum of Agreement was signed by all interested parties committing them to the establishment of a multi-disciplinary project team with the primary aim of preparing a Plan of Action. The respective Plans of Actions for each of the 12 forests was submitted to and approved by the Government in 1993. In addition to a Plan of Action, based on extensive public consultation, a detailed cost–benefit analysis was prepared by the Forestry Commission to Treasury rules. The Government's consideration of these plans was both in terms of the specific proposals they contained as well as the financial implications for public expenditure.

Whilst these plans were primarily designed to respond to local circumstances, there was an expectation that they would have a common structure and achieve a broad range of objectives. The impact of these plans could be quite considerable. Over half of the population of England live within or have easy access to Community Forests. Such coverage makes them the largest single environmental initiative that has been launched for many years. Arguably the programme is of a scale similar to that created for the National Parks or Areas of Outstanding Natural Beauty.

Approaches to Plan Making

Basically there are two main approaches to plan making. The first option is a top to bottom approach. This means that a master plan or grand vision is produced which is then imposed on the designated area. Good examples of this type of approach are the Greater London Development Plan produced by Abercrombie and the various master plans produced for new towns. Both of these are good examples of successfully applying a top to bottom approach. In the case of London, the restructuring required following the devastation of the war required a grand vision to be produced which could lead the nation's recovery.

The Abercrombie plan provided the green lungs that were much needed by Londoners to relieve the urban sprawl that had developed in the interval period. In view of the circumstances the nation found itself in during the early 1940s an incremental approach to plan making would not have been successful. Similarly the rapid urban expansion immediately after the war created an immediate demand for new housing. Again circumstances dictated that a top to bottom approach to plan making would be the most appropriate.

Many of the locations chosen for the proposed new towns were open countryside with very few inhabitants. This is very true of Harlow New Town where apart from a few hamlets the area was mainly agricultural. This enabled the architect planner, Sir Frederick Gibbard, to come up with a fully integrated plan that created a brand new environment for Londoners which contained vast areas of public open space, an effective road network and wide range of facilities. The development plan provided an integrity to the large-scale development that was occurring which would have otherwise been missing if an incremental approach had been adopted. It ensured that the provision of housing was matched with the provision of adequate public open space, social facilities and the necessary infrastructure of a growing town.

There are, however, circumstances where the imposition of a grand design as used in London and the new towns, would be entirely inappropriate. This is particularly true where there are established communities or habitats that have value and need to be taken into account. It must also be recognized over four decades on from the greater London Development Plan and the various new town master plans, that society has changed significantly and in particular the role of the planner has changed. In the immediate post-war period and certainly into the 1960s and 1970s it was accepted practice for the planner, as the 'expert', to provide all the solutions for local communities.

This role, however, became increasingly under pressure from the mid-1970s onwards, when the need to restructure our urban areas became pressing. Local communities quite rightly demanded an input into the plan-making process. This pressure changed the role of planners. No longer providers, they increasingly became enablers and facilitators. In such a role the approach to plan making itself had to change. It was no longer appropriate in such circumstances to come up with blueprints or master plans in isolation. Such plans had to be developed with local communities and the plan-making process had to be responsive to their views. This suggested that the plan-making process would need to be more incremental, starting with local issues and circumstances and gradually building up to an overall plan of action. This bottom to top approach of plan making offers far more scope to involve local people in the plan-making process. It is important, however, to recognize that this approach to planning is not a new one but was pioneered at the beginning of the century by planners such as Sir Patrick Geddes.

Often referred to as the father of town planning, Sir Patrick Geddes offered an incremental approach to planning that was based on the principles of survey, analysis and then prescription. He provided a direct contrast to many Victorian and Edwardian planners who thought in grand terms and produced elaborate master plans. Coincidentally, Sir Patrick Geddes was one of the first planners to use the new practice of photography to survey the sites and areas that he was investigating.

Plan Making in the Forest of Mercia

The designated area for the proposed Forest of Mercia was very unlike that chosen for the new town of Harlow. A long-standing area of mining and industrial activity, as well as farming, it has a resident population of 250,000. It is also situated at the point where the southern landscapes merge with their northern counterparts and consequently produces quite extensive landscape variety. It is therefore a very complex area and one that could not easily be accommodated within an overall master plan produced in isolation.

The local partnership set up to create the Forest of Mercia placed a great deal of emphasis on involving local people in the plan-making process and therefore also rejected the idea of the top to bottom plan-making process. Adequate time was therefore set aside to ensure that local people could if they wish become actively involved in the process or, as in the case of the vast majority, be kept informed of progress and have a chance to comment on the work being done. To ensure that such a situation arose, it was decided that a three-stage approach would be adopted. These three stages were based on Geddes' survey analysis and prescription.

To facilitate public involvement, a wide range of interest groups and organizations were invited to participate in an advisory panel set up to guide the plan-making process. The groups attending this panel varied greatly and included representatives of the land-owning and farming communities, such as the Country Landowners Association and the National Farmers' Union, to organizations committed to support public use of the countryside, such as the Ramblers Association and the British Horse Society. These groups had the opportunity both to participate in the plan-making process as well as the practical creation of the forest by becoming involved in tree planting and woodland management projects.

It was agreed that each stage of the plan-making process would have a report which would be considered by all interested parties prior to the next stage beginning. It was recognized that it would prove difficult to involve individual members of the community outside of these groups. The resources therefore were primarily focused on ensuring representative groups were kept fully involved although any member of the public wanting to participate could do so. An extensive programme of press releases and promotional events was undertaken to provide as much information to the public about the work that was being undertaken.

The plan-making process pioneered by Geddes is an incremental one with each stage leading into the next. The benefit of this approach is that it allows genuine participation and the opportunity for decisions to be influenced by local people. It also means that everybody is aware of the issues, has the opportunity to comment on them and contribute to the prescriptions that were brought forward. By utilizing such an approach the proposals that were eventually contained in the Forest Plan were not a surprise to anyone. They had seen the local circumstances, the issues that existed and understood why particular proposals had been brought forward.

This approach can be illustrated by the following example: a local farmer is under pressure from a local amenity group to safeguard a stand of oaks that are significant features in the landscape, but at the time are also entering old age and posing constraints on the agricultural activities that are being undertaken. The farmer would ideally like to remove the trees whilst the amenity group are keen to safeguard a cherished local feature. This is not an uncommon scenario. With Geddes approach to planning, the following processes would be undertaken:

1. A full survey would be undertaken of the land in question. This would avoid any value judgements about the features or components of the landscape. It would really record as a matter of fact the number of trees, their condition, age and features. No judgements about the value or advantages of the trees would be made at this stage.
2. This stage would in effect be an analysis of the value of these trees. It would look at both the opportunities and constraints associated with the trees and the adjoining land. Very often a SWOT (strengths, weaknesses, opportunities and threats) analysis will be undertaken to determine their value. In this case the trees are oak trees which have a high conservation value. They are all in an advanced stage of age with some disease associated with their age, that they restrict the agricultural use of the land because of their position within the field and that they are indeed significant landscape features. Even at this stage no views are put forward as to what should be done with the trees or the land. It is as the title implies, an analysis of the options that relate to the site.
3. Prescription. It is at this point, once agreement has been reached on the extent or character of the site, a description of the various components of the site and an analysis of the opportunities and constraints that exist will then be possible to bring forward any suggestions on how the perceived problems can be overcome. In the example that has been used, the prescription that may emerge could be along the lines that as the trees form an important landscape and conservation asset within the area, their removal now would

be undesirable. However, in view of their age, the existence of some disease and their constraint on agricultural activity, inevitably they will need to be felled. To plan for such circumstances the prescription that would be offered would be as follows:

1. New trees of the same species of provenance would be planted now. These trees would be planted in the hedgerow and field corners to avoid restraining agricultural use of the remainder of the site. They would be planted in readiness for the loss of the trees through old age or disease in the next 15–25 years.
2. The existing trees would receive remedial tree surgery to deal with any disease and prolong their life expectancy.

Although this is a fairly simple example, it does illustrate the process involved and conveys the potential value of such an approach in problem solving. It is a particularly useful approach when there is acknowledged conflict within an area and the need to find a consensus.

As a result of adopting this approach, the plan that was produced for the Forest of Mercia was submitted to Government without objections. As indicated, a wide range of organizations, often representing very different interest groups, were able to participate in the process and find common ground. The plan has also provided a very positive framework within which implementation has been progressed. This inclusive approach has meant that there is no opposition to the landscape changes that are being brought about.

To ensure that the plan remained responsive to local circumstances and expectation, a 5-yearly review has been built into the programme. The plan covers a 50-year period and clearly in view of the changes that have taken place in the last 50 years, it is very unlikely that today's thoughts will necessarily be the same as in 30 or 40 years time. The review will be carried out on the same basis as the production of the plan itself to ensure that local people have the opportunity to participate and feel included in the process.

Wider Application of the Approach

In addition to utilizing this approach to produce the Forest Plan, the same methodology is being used on an European-funded project looking at the potential of renewable energy in the Forest area. The European Commission has provided funds under the ALTENER Programme for the Forest of Mercia to work in conjunction with the Dutch Forestry Service. The project is designed to explore the potential of wood as a renewable source of energy. Again, as the plan will be dealing with existing activities and resources, it was felt that the approach offered by Geddes would provide a much more suitable basis to preparing the plan. The survey stage will be looking at both the resource of timber as well as current energy use and the technologies available to produce energy from wood fuel sources. One of the key outputs from the work will be guidelines that could be utilized by other local authorities in the European Union to produce their own local plans assessing the potential for developing renewable energy. The principle of a bottom to top approach to plan making has been accepted by the Commission and this will form the basis of the guidelines. Again it is felt to be the most suitable way of including the business community and energy users in general in this important piece of work.

Conclusions

Clearly there are different ways of preparing plans and this paper has focused on the two main methods. Both approaches are quite legitimate, but it is apparent that their suitability is determined by the objectives to determine the need for the plan as well as local circumstances. It is highly unlikely that had a top-down approach been adopted that a plan would have been produced that would have had such widespread support. Inevitably, groups within the community would have felt excluded and threatened by the plan. Choosing the most appropriate plan-making process has also ensured that the implementation stage has enjoyed full public support. Increasingly all groups within the community are playing a role in the implementation of the Forest as they see that their views have been taken into account, that reasonable courses of action have been put forward and that they are given a continuous opportunity to participate and comment.

References

Forest of Mercia (1993) The Forest of Mercia Forest Plan, October 1993, Cannock (England): Forest of Mercia.

Forest of Mercia (1998) Chronicles of Mercia: Annual Report & Business Plan 1998/99, Cannock (England): Forest of Mercia.

Forest Tourism and Recreation in Nepal

14

Trevor H.B. Sofield

Introduction

The Hindu Kingdom of Nepal, located between India and Tibet, is a small country of only 151,000 km^2. From south to north it is less than 180 km wide but it encompasses a very diverse biogeography, ranging along a south–north transect from tropical lowland forests and grasslands of the flat *Terai* to the alpine tundra of the massive Himalayan ranges with the world's highest peak, Mount Everest (8848 metres or 29,028 feet). The hot southern forests of the Chitwan hills and plains are the habitat of the endangered Bengal tiger, the single-horn Asian rhinoceros and the gharial crocodile, while the northern alpine forests and peaks are home to the even more highly endangered snow leopard and the red panda, among others. More than 500 species of birds have been recorded in Nepal, over 480 species in the *Terai* (Mishra and Jeffries, 1991).

The monsoonal system and altitude are the two major determinants of vegetation zones in Nepal. Thus, East Nepal receives the full benefits of the wet winds from the Bay of Bengal and rainfall in many regions exceeds 3000 mm (120 inches) per year. As the monsoon proceeds westward rainfall decreases, and in the north-west the high Himalayas block the monsoon and form a rainshadow where less than 300 mm (12 inches) is received. This east–west gradation is cross-cut by a number of altitudinally affected climatic zones. The tropical zone extends up to about 1000 m (3300 ft). The vegetation is quite uniform throughout the east–west length of Nepal despite variations in rainfall. The dominant tree is *Shorea robusta*, a hardwood tree valued for its timber, and this type of forest is called sal forest after its common name. It grows in association with many other species such as

Dillennia pentagyna (whose fruit is used by villagers for making pickles), *Semecarpus anacardium* (called the nut-marking tree because the dark juice of its fruit is used as ink, and also as an abortificant); and *Spatholobus roxburgii*, an enormous vine which grows clockwise around its hosts (folklore has it that if you find one growing anti-clockwise a nail hammered into it will turn to gold).

The sub-tropical zone extends between 1000 and 2000 m (6600 ft) and in the eastern and central regions is dominated by two evergreens, *Schima wallichii* and three species of *Castanopsis* (chestnuts). In the drier west, the dominant species is *Pinus roxburghii* (chir pine, whose resin is valued for making rosin and turpentine).

The warm temperate zone between 2000 and 3000 metres (9900 ft) is often referred to as laurel forest in eastern and central Nepal because of the abundance of *Lauracae*. Rhododendrons (the national flower) are found from around 1100 metres and their range extends up to 2500 m, often forming the understorey in laurel forests. There are regions where they grow almost exclusively, forming a canopy 15 m (50 ft) high, and these now constitute a focal tourist attraction. In the drier eastern regions the temperate zone is dominated by oak (*Quercus incana*), conifer, cedar and birch. Conifers include the low altitude fir, *Abies pindrow*, which grows to 45 m (150 ft), the silver fir, *Abies spectablis*, and the blue pine, *Pinus exclesa*. The Himalayan cedar, *Cedrus deodara*, forms large forests along the Karnali River system.

The cool temperate or sub-alpine zone ends at the tree line at about 4000 m (13,000 ft). The tree line in much of Nepal is defined by the birch, *Betula utilis*, and its papery bark is used for a number of purposes such as a lining for ceilings and as a natural 'grease-proof' paper for wrapping yak butter. The zone above 5500 m (18,000 ft), the aeolian or nival zone, is effectively a windswept region of rock, snow and ice, devoid of most vegetation except for the hardiest of lichens.

Traditional Forest Use for Recreation in Nepal

In an historical context Nepal has had a forest protection regime for centuries. There are three main elements to this: royal forest reserves, sacred sites' forests and community forests.

Royal forest reserves

The rulers of Nepal have for centuries set aside reserves for hunting by royalty. Some of the earliest records of royal forest reserves extend back to the era of the Malla dynasty which ruled Nepal from 1200 to 1769 AD. The Shah dynasty which replaced the Malla kings after a battle in that year continued the practice. The rainforests of Chitwan, for example, were preserves for hunting tigers and rhinos by King Prithvi Narayan Shah and his descendents for centuries before being gazetted as a national park. The last of the now infamous tiger shoots by the then king of Nepal in 1937 bagged more than 100 tigers and rhinoceroses (Allen, 1999).

Some forests were preserved for royal retreats, one of the most famous being that of the palace of Ranighat Mahal, situated in dense forest on the banks of the Kaligandaki River 7 k from Tansen, once the main town on the trading route between Tibet and India. This palace, originally surrounded by extensive pleasure gardens, was built by one of the Shah kings for his wife some 200 years ago and is the Nepali equivalent of India's Taj Mahal.

Where in the past forest reserves had been set aside for the recreational pursuits of royalty, under the impact of contemporary tourism, community forests and protected areas are now becoming recreational resources for local residents and tourists. For example, on the Shreenagar plateau above Tansen the 200-ha forest reserve was used exclusively by the ruling elite for archery, then shooting and hunting and horse-riding before its present status as a people's park. It is now visited by several hundred picnicking Nepalis every weekend, and the numbers increase to several thousands when special events are staged there. Its extensive stands of red rhodendron forests are a feature. It boasts a nine-hole golf course, meditation spots and camping facilities. In a novel recreational twist, the Shreenagar Park has been the site for more than ten full-length movies in the past decade, because the Himalayas from Dhaulagiri and Annapurna in the west to Manaslu and Ganesh Himal in the east form a spectacular backdrop of snow-capped peaks. Shreenagar forest was once private space and has now become public space, and the elitist social space of the ancient courts has given way to leisure space for families.

Religious forests

It has been said that Nepal has more temples and gompas (monasteries) per capita than any other country in the world, and the religious devotion of its Buddhist and Hindu populations over the centuries has resulted in thousands of sacred sites. It is often difficult to distinguish between the two main spiritual practices of Nepal. The highest peaks of the Himalayas have been worshipped by animists, Buddhists and Hindus as the 'centre of the universe' and all of the higher peaks are sacred.

Trees and forests have been an integral component of virtually all sacred sites and have provided a range of recreational facilities for pilgrims and other visitors for centuries (including shelter, food and toiletting). Hilltops have been a favoured site for temples and their forested slopes have been classified as sacred, responsibility for their conservation and protection vested in the custodianship of resident communities, and monks. In the now relatively densely populated Kathmandu Valley, the sacred forests of ancient hilltop temples such as Svayambunath, Changu Narayan and Pashputinath (all World Heritage sites) are oases for wildlife. They provide a safe environment for rhesus macaques, the mongoose, lizards, a large variety of birds and insects, etc., despite the many hundreds of thousands of pilgrims and tourists who throng these heritage sites. Individual trees may also be declared sacred and pipal trees throughout Nepal are accorded this status, based on the belief that Buddha gained enlightenment under such a tree while meditating near Benares in India. Religious beliefs have thus played a significant role in the protection of many thousands of small forests throughout Nepal for centuries (Ingles, 1990).

Community forests

Traditionally many forests in Nepal were managed by communities which made arrangements to protect and regulate access to forest resources for which there was no single owner. Gilmour and Fisher (1992: 40) have termed such systems of protection, regulated access, utilization and distribution of forest products as 'indigenous forest management systems'. Anthropologists have recorded a range of sustainable forestry practices with community-based sanctions to ensure the perpetual availability of forest resources (e.g. Molnar, 1981; Messerschmidt, 1987; Campbell *et al.*, 1987; Gautam, 1987). In rural Nepal, people who spend several hours a day collecting products from trees clearly recognize their importance. Indigenous technology such as lopping of lower branches rather than cutting the main stem, restricting firewood to collection of dead material, and constant if irregular replanting from both cuttings and seeds (even if not in the serried rows of western-style plantations), accompanied by village management systems such as the *shinga naua* system of the Sherpas of Solu-Khumbu (Furer-Haimendorf, 1964, 1984) attest to this understanding. The *shinga naua* are appointed by local Sherpa communities for fixed terms (several years maximum) as guardians of common forests with responsibility for allocating forest resources and making sure that villagers adhere to community-determined rules for usage.

While the focus of indigenous management has been on resource conservation and utilization, community forest products were also associated with a variety of recreational uses. For example, many of the flowers were gathered for decorative purposes, and for inclusion in *pujas* (offerings to the gods). A particular tree in the subtropical zone (*Mallotus philippenensis*) provides the red powder from its fruit which is traditionally used as the ceremonial red dust in Hindu rituals. People of the alpine zone have for centuries valued the two endemic dwarf rhododendrons *R. cowanlum* and *R. lowndesii*, for incense and as a substitute for snuff. The bark and fibres of other species of trees and plants are used for making yarn for cloth, paper and bedding, and a wide variety of medicinal herbs, ferns and lichens are gathered from the forests. Many of the larger community forests have also been utilized over the centuries for hunting, although the recreational element of trophy hunting has been subordinated to hunting for more practical purposes such as meat and hides. The musk deer (*Moschus moschiferous*) is still hunted – despite being protected – because the musk, secreted by the male from a pod under its tail, is a valued and highly invaluable ingredient in Chinese medicine, Tibetan incense and the western perfume industry (Ramble, 1999).

As tourism has developed into a major industry in Nepal over the past 30 years, community forests are now being re-planted, protected and utilized for recreational purposes, and this aspect is explored in further detail below.

Contemporary Protected Area Management

Despite its extreme poverty – Nepal is one of the world's 25 least developed countries and its population of more than 22 million has a per capita annual income of less than US$100 – it has one of the better forestry protection regimes in Asia. Its protected areas now total more than 14% (21,000 km^2) of the country, and consist of eight national parks, four wildlife reserves, two conservation areas and one hunting reserve (Yonzon, 1997). Two of those national parks, the Royal Chitwan National Park and the Sagarmatha (Everest) National Park, have been accorded World Heritage Site Listing because of their outstanding natural values. Legislation supporting protected areas commenced with national parks in 1973.

Nepal's protected areas were established with the twin objectives of safeguarding biodiversity and maximizing tourism benefits for the country. There is a growing alliance between protected areas, forests and tourism, with more than 175,000 international visitors touring Nepal's parks in 1997/98 (Department of National Parks and Wildlife Conservation (DNPWC), 1998). Tourism-related activities in protected areas contributed more than 90% of total revenue generated for the national parks system in 1996 (Yonzon, 1997). Four protected areas receive more than 95% of all national parks visitation – Chitwan, Annapurna, Sagarmatha and Langtang, the latter three being high mountain trekking areas (Banskota and Sharma, 1995a, 1995b).

Forestry legislation pre-dated that of protected areas with the Private Forest Nationalization Act of 1957. One of its major aims was to place all forest land under the control of the Forest Department to 'prevent the destruction of forest wealth and to ensure the adequate protection, maintenance, and use of privately owned forests' (Regmi, 1978: 348). However, little more than token action resulted because there were only five or six trained foresters in Nepal at the time (HMGN, 1986). The Forest Act, 1961, provided for land to be made available for small private forest plots and introduced the idea of transferring government forest land appropriated under the

1957 Act back to village community political units (*panchayats*). There was little change on the ground nevertheless since the Forest Department still held regulatory authority and policed forests with assistance from the Royal Nepal Army (which is stationed in every national park except the Annapurna Conservation Area). Following a political revolution in 1990, the abolition of the *panchayat* system, its replacement with Village Development Committees throughout Nepal, and the Forest Act 1993, which established Forest User Groups which could apply to regain control of community forests, substantive changes occurred. Since 1993 control of several hundred small forests has been transferred to communities, and these are known as Community Forests. They usually involve a relationship between villagers and foresters (similar to that developed under the Annapurna Conservation Area Programme – ACAP), through which the partnership fosters an enhanced capability by villagers to manage their forests better both for access to its products and for conservation and transformation to better quality forests.

Challenging Orthodoxy

In examining Nepal's approach to the conservation and utilization of its forest resources, two popular assumptions are challenged. The first is the so-called 'deforestation crisis', supported by what Ives and Messerli (1989) have termed the 'Theory of Himalayan Environmental Degradation' in their critical appraisal of the overstatement of the crisis. The second is the 'Tragedy of the Commons' (Hardin, 1968) which has its counterpart in Nepal. After exploring these two issues, two institutionalized examples of good practice utilizing sustainable approaches to the environmental management of forests for tourism and recreation will be reviewed. These are the Annapurna Conservation Area Programme (ACAP) and the Parks and People Programme (PPP) of the United Nations Development Programme.

Theory of Himalayan Environmental Degradation

The 'Theory of Himalayan Environmental Degradation' links high population growth and increased tourism consumption of forest resources in the Middle Hills and high mountains to unsustainable practices which are, according to the claimants, resulting in an ever-widening circle of denuded hillsides around each village, more frequent landslides because of the depletion of ground-holding trees, an ever-increasing loss of topsoil because of increased erosion, and a greater incidence of flooding by swollen rivers because of the deforestation (Eckholm, 1975). In its more extreme form the frequent and disastrous floods of the Ganges River and its Delta in India and Bangladesh far downstream have been blamed on deforestation in Nepal. This theory was reinforced by reports such as that issued by the World Bank (1978) which suggested that the hill areas of Nepal would be totally deforested by 1993 (i.e. after 15 years) at the then estimated rate of depletion, and the Terai would face a similar fate by 2003. The acceptance and dissemination of this theory has been so widespread, particularly its repetition in the tourism and recreation literature where it has had the status of self-perpetuating fact, that as the year 2000 approaches most visitors to Nepal expect to see bare hills and mountains (and are surprised at the extent of forest cover).

During the past 15 years numerous researchers have demonstrated that the theory is flawed in at least three key areas. Firstly, deforestation in the Middle Hills and high mountains is not as widespread as claimed nor is it increasing. Secondly, deforestation is not a recent phenomenon which has occurred largely in the past 40 years because of population pressure (Ives and Messerli, 1989; Gilmour and Fisher, 1992). Thirdly, tourism is not a major contributor to the alleged deforestation of the high mountains. It is incontestable that many of Nepal's forests are in a degraded state but the magnitude of the crisis propounded by the theory which projected an image in the 1970s of bare hills and mountains by the 1990s is clearly not accurate as we approach the end of the millennium.

With reference to the first assumption of the theory, that there has been massive deforestation of the Middle Hills and mountains, a thorough investigation of Nepal's forestry base was undertaken as part of the Government's Land Resources Mapping Project (LRMP) in the early 1980s. The LRMP utilized aerial photography supported by detailed on-ground surveys. It established that 38% of the country was covered in forests – 46% in the hills and mountains where the 'crisis' was considered to be most severe, and about 34% in the Terai. A 1983 Nepalese Government survey which measured forest cover through a comparison of aerial photography taken in 1964 and 1978 discerned that there was in fact a slight increase in hill forests during this period. A subsequent Nepal–Australian Forestry Project substantiated the fact that forest cover remained at almost 40% for the entire country (Gilmour, 1988; Gilmour and Fisher 1992). Site-specific studies by other researchers demonstrate that in some instances forests covered more than 60% of a hill or mountain district, that not only had forests in such sites been maintained in the past 40 years but reforestation had expanded forest cover, and tree density had increased under initiatives by local communities (e.g. Bajracharya, 1983; Mahat *et al.*, 1986a, b; Griffin, 1987; Messerschmidt, 1987; Fisher *et al.*, 1989; Gilmour and Fisher, 1992; Gurung, 1995; Sofield, field notes on Sirubari, 1999). In the Annapurna Conservation Area, which covers 7629 km^2, forest cover has increased over the past 15 years by more than 20% under regeneration projects supported by village communities, and under reforestation projects on both government and private lands more than 1 million saplings have been planted to create more than 600 ha (6 km^2) of plantations (ACAP, 1998). Rather than Eckholm's prediction of villages surrounded by a widening circle of denuded hillsides the reality is often the complete opposite (Gilmour and Fisher, 1992: 33).

The second assumption of the theory, that deforestation is a recent phenomenon of the last 40 years due to population pressures and tourism, is also disputed by recent studies. Gilmour and Fisher (1992) demonstrated, for example, that most of the land converted from forests to terraces and agriculture in the Middle Hills took place 80–100 years ago when the taxation policies of the Rana regime provided incentives to do so. In the same way, much of the high mountains have maintained their current ratio of agricultural lands to forests for the past 80–100 years. Population pressures in some areas have resulted in decreases in forest density but Gilmour and Fisher (1992: 33) suggest that in at least as many other areas 'the forests are now in much better condition, in terms of tree density, height and amount of regeneration, than they were in the recent past', largely because of village initiatives to protect their resources.

However, in other areas there has been over-utilization of forest resources. Yonzon (1997), for example, reports that for the high mountain area of Upper Mustang (2567 km^2) where there has been a heavy influx of Tibetan refugees, by 1997 only 16 km^2 of forest and alpine shrubland remained out of the district's total of about 20 km^2 in 1960. This district lies in a rain shadow and extreme altitudes mean that more than 90% of the land is alpine tundra, alpine grasslands, barren rock or permanent snow and ice. Except for a few valley floors the so-called arable land is marginal and itinerant grazing (sheep, goats, yaks, cattle) rather than agriculture is the norm. Even here, the situation is changing and in the past 3 years, re-afforestation efforts under the guidance of the ACAP have increased forest cover by an additional 10% (Gurung, C., personal correspondence, March 1999).

In the Sagarmatha (Everest) National Park (SNP), mixed impacts of tourism on forestry resources have been observed. Initial tourism into the region resulted in substantial degradation of the forests but in the past 20 years there has been significant improvement. Since 1976 the SNP has enforced a strict ban on the cutting of live wood, except where a permit has been issued for construction purposes (Rogers and Aitchison, 1998). Secondly, all group trekkers are required to use kerosene for fuel. Thirdly, the installation of a 650 kW hydro-electricity scheme at Thame has significantly reduced fuelwood consumption. A survey between 1993 and 1996 indicated that lodges had halved their consumption of fuelwood and households had reduced their consumption by two-thirds (Rogers and Aitchison, 1998). In addition, substantial effort has been put into local involvement in forest management by the Himalayan Trust (especially by the re-introduction and employment of *shinga naua* – traditional Sherpa forest guardians) and this participation has been fundamental in ensuring that SNP forests are now generally well-maintained (Baker, 1995; Ledgard, 1997). However, Pharak Forest, which serves as a transit corridor into the SNP, faces a continuing high demand for fuelwood and construction timber; it has experienced difficulties with community participation and its management is problematic (Bauer, 1995). It constitutes a localized crisis in terms of sustainable practices, although Rogers and Aitchison (1998) report that increased community involvement (expansion of community user groups from two to four) has begun to improve the situation.

In the lowland Terai, the situation is different. There is clear evidence of recent and extensive loss of forests, the Government survey of 1983 indicating that between 1964 and 1978 the area of Terai forest declined by 191,000 ha to 593,000 ha, or almost 25% (LRMP, 1986). The Terai witnessed a huge increase in population following the eradication of malaria in the 1950s which permitted permanent habitation, with re-settlement from the Middle Hills and migration from northern India. In 1952, the population of the Terai was about 2.9 million, by 1981 it was 7.03 million, and by 1991 it was more than 10 million (Goldstein *et al.*, 1983; National Planning Commission, 1998). Much of the central and eastern Terai is now under cultivation where 40 years ago it was heavily forested (Gilmour and Fisher, 1992).

However, the theory of Himalayan environmental degradation links deforestation of the hills and mountains to increased erosion, sedimentation and flooding both locally and downstream in India and Nepal, and *not* to deforestation of the flat lowlands of the Terai. A comprehensive review by Bruijnzeel (1989) concluded that the young geological age of the Himalayas and its accompanying torrential rains were more important factors in landslides, erosion and resultant hydrological impacts than human intervention. He found that stream flow in smaller catchment areas of less than 500 km^2 could be affected by vegetation and land-use practices but this effect disappeared as larger areas were considered; that depending upon infiltration characteristics of the soil, conversion of forest land to agricultural land could lead to either an increase or a decrease in seasonal river flows; and that with reference to widespread flooding, especially far downstream, 'the presence or absence of a forest cover is of negligible importance in ... the circumstances of very high rainfall prevalent in the Himalayan foothills' (Bruijnzeel, 1989: 118). Ives and Messerli (1989) reached a similar conclusion: human intervention could be critical at the micro-watershed level or individual mountain slope but even then periodic catastrophic rains tended to override the effects of human activity. Virtually all forestry workers in Nepal currently can provide anecdotal evidence of mature forests where there has been neither logging, agricultural encroachment or roadworks undercutting slopes which yet suffer from major landslides every wet season.

With reference to the third assumption of the theory, that increased tourism has contributed in a major way to deforestation, much of the leisure and recreation literature has elevated the theory to the point of undisputed fact. Fisher (1990: 69) suggested that the theory had become so pervasive as to take on the status of a myth, in the anthropological sense that it provokes a needed response to counter it, in this case 'a charter of action, justifying a vast amount of foreign aid into reforestation and watershed management projects'. It has also been mythologized in the more mundane sense of the word – 'it is a popular, but unjustified, belief' (Fisher, 1990: 69).

Studies by ACAP scientists, foresters and others indicate that the use by villagers of leaf material as fodder for their animals is up to 7.5 times greater than for woody material as fuel biomasss (Mahat *et al.*, 1987), even when tourist consumption is added. Banskota and Sharma (1996) have attempted to measure consumption of fuelwood per tourist. Others (e.g. Thompson *et al.*, 1986) point out that there are so many variables associated with consumption rates of different forest resources that such figures should be treated with caution. Sofield and Bhandari (1998) observed that in the Langtang National Park trekking villages, little firewood is gathered for tourist-associated consumption (cooking in the lodges of Thulo Syabru and Syabru Besi, for example, is by kerosene) although villagers are active in stockpiling wood for heating in winter (when there are no tourists).

Of more than 20 current tourism development projects being carried out in Nepal by a variety of international non-governmental organizations (INGOs) which were reviewed in 1998, every one of them without exception included specific components to counter fuelwood consumption, e.g. the UNDP's Quality Tourism Project, the Dutch aid agency SNV's village development programme, The Mountain Institute's Langtang National Park Ecotourism Project, etc. (Sofield and Bhandari, 1998). Most of them, for example, have promoted the use of kerosene, to the extent that the Government itself now requires all organised trekking groups to carry their own kerosene needs for cooking. ACAP has established some 21 Community Kerosene Depot Management Committees in the villages along the major trekking routes in the Annapurna Ranges (Yonzon, 1997). Other initiatives include the provision of back boilers for cooking, solar panels for hot water, and re-afforestation schemes. In some instances (e.g. Ghandruk and Sikles villages in the Annapurna Conservation Area), micro-hydro schemes have been installed to decrease dependence upon forests for fuelwood. Where the 1970s mountain tourism appeared to be having an adverse impact on forests, the 1990s mountain trekking scene is one where villages on the main trekking routes have increased their forests, plantations and orchards, and appropriate technology is contributing to sustainable management of demands on the forest resources. This is despite the fact that trekking numbers have increased from 25,000 (approx.) in 1970 to more than 92,000 in 1998 (Ministry of Tourism & Civil Aviation Annual Report, 1998).

Ghandruk provides a specific case in point: in 1970 it received less than 5000 tourists, and its hills were severely denuded. In 1998 it received about 31,000 trekkers and it is now surrounded by more than 300 ha of forests, both naturally regenerated and planted, because of the local village community's active participation in forest management with guidance from the ACAP.

In summary, it is suggested that the 1970s notion of a severely denuded Himalayan Nepal with an alarmingly high deforestation rate and resultant magnitudes of erosion and environmental degradation was flawed. In the 1990s a more balanced view is that a crisis of the proportions envisioned was over-stated – although the responses provoked from the Nepal

Government, international aid agencies and local communities to the perception of a crisis may have led to a mitigation of the problem since all have been actively engaged in re-afforestation and regeneration efforts.

Community forests in Nepal and the 'tragedy of the commons'

A Nepalese version of Hardin's (1968) 'tragedy of the commons' was linked to the deforestation theory outlined above, with the prevailing view of the 1960s and 1970s that rural communities were, because of their poverty and ignorance, responsible for much of the degradation of the country's forests (e.g. Rieger, 1981). This is a view which has been strongly contested by others (e.g. Hobley, 1985; Gilmour and Fisher 1992). Traditionally many forests in Nepal were common property and available for use by communities (as noted above), and the Nepalese version of the 'tragedy of the commons' was that because of increasing population everybody simply took what they needed from those resources not owned by anybody in particular: the assumption was that inevitably common property resources would be over-utilized. Essential to this view was also the assumption that rural populations were ignorant about the value of forests and needed to be educated about the importance of trees. Legislation in the 1970s therefore enabled the government to appropriate common forests throughout Nepal so that trained technicians could 'correct' the situation and reverse the assumed degradation.

Anthropological literature vigorously refutes this last assumption that local people are ignorant of good forestry practice. Dove and Rao, in considering indigenous forestry practices in Pakistan (1986), accept that Hardin's theory of the inevitability of over-use of community resources *in the absence of sanctions* is accurate as a general proposition, but that his view that sanctions are rare is incorrect empirically. Bromley (1986) in a paper prepared for the International Centre for Integrated Mountain Development (ICIMOD), suggests that there is a need to distinguish between 'open access' (resources available to everyone) and 'common property' (resources for which specified people have specified rights) and criticizes the simplicity of Hardin's notion of 'common resources' because it fails to draw this distinction. In examining the situation in Nepal it is obvious that there are many communities throughout the country where there have been long-standing local management systems designed to protect forests, regulate use and impose sanctions where over-exploitation occurs (Campbell *et al.*, 1987; Messerschmidt, 1987; Gilmour and Fisher, 1992). Furer-Haimendorf (1984) has argued that the replacement of these systems with Government Forest Department supervision, often ineffective, paradoxically created the very situation they were intended to deter, i.e. Hardin's classic case of over-use. No longer responsible for the continued productivity of the forests, the local people have attempted to exploit the resources to maximum extent.

In this context the creation of Nepal's network of protected areas and forests has provoked significant conflict between park managers and regulatory forces on the one hand and local communities on the other. In many cases agricultural communities have been re-located outside park boundaries, or excluded from utilizing resources to which they traditionally have had access. They are not interested in protecting the parks and reserves as they see them as a free resource to be exploited for their benefit, they are by and large excluded from the benefits of tourism to the protected areas, and they see park staff as a barrier to their pursuit of basic needs (UNDP, 1994). Illegal logging, illegal gathering of fuelwood, fodder and thatching grass, illegal grazing of livestock, poaching, and killing of animals because of crop damage, cattle depredation and human injuries, and in some instances competition for water resources, pose significant threats to the conservation and biodiversity of the protected areas by adjacent communities.

Carter and Gilmour (1989) documented a different response in two Middle Hills districts where common forests were depleted: extensive planting of trees on private lands by farmers, without government direction, to meet their needs. Between 1964 to 1988, tree density per hectare in four sites which they surveyed increased on average by 233 trees per hectare or 357%, clearly a flourishing agro-forestry system on lands directly under the control of local farmers. This phenomenon has been one of the least researched in the literature on forestry in Nepal, perhaps because the Himalayan degradation theory and the 'commons tragedy' notion of inevitability of over-exploitation both assumed the existence of a vacuum of knowledge by local communities for which the 'charter for action' demanded extension services and education of farmers by trained foresters. Until the last decade, evidence of dynamic and productive private silviculture by Nepalese villagers, while noted in passing, tended to be ignored. The pioneering work of the ACAP has been influential in changing this view and accepting that rural communities in Nepal are capable of effective forest management. This is not to state that all local forest management activities in Nepal have been competent in terms of sustainable practices, but rather that the extremes of two commonly held stereotypes – that peasant farmers are ignorant and incapable of sensible resource management without education and its converse, that the inherited wisdom of centuries has all the answers if only communities were left to their own devices – are false, as is the case with many stereotypes. As Gilmour and Fisher (1992: 56) noted, 'villagers are not all ignorant, nor are they all wise'.

Annapurna Conservation Area Project (ACAP)

The Annapurna region contains some of the world's highest and most beautiful peaks (Annapurna 1 rises to 8091 m) and in 1997 more than 60,000 trekkers, accompanied by approximately 40,000 support staff (guides, porters, cooks, etc.) walked its trails. The Kali Gandaki River, which forms one of the world's deepest valleys, flows from its heights to its southern lowlands. The Annapurna Conservation Area (ACA) encompasses two distinct climatic regions, the southern districts receiving the country's highest annual rainfall of more than 3000 mm while the northern trans-Himalayan districts receive only 250 mm per annum. Such variations in climate and geography support a wide range of habitats with significant biodiversity. There are more than 100 species of animals including the snow leopard, musk deer, blue sheep, Himalayan thar, red panda, and the only area where the brown bear, Tibetan argali, all six species of Himalayan pheasant and the black-necked crane are found in Nepal, and more than 1200 species of plants. It supports an ethnically diverse population of more than 120,000, most of them subsistence farmers, who live in 55 village districts. It was the first protected area in Nepal to allow local residents to live within its boundaries and maintain their traditional rights to access and use of its natural resources. It was also the first protected area managed by local experts that did not use the assistance of the army to protect the resource base on which the region depends (Ghurmi, 1997).

The ACAP began as a pilot project run by the King Mahendra Trust for Nature Conservation (KMTNC) with the Ghandruk Village Development Committee (VDC) in 1986. This project was designed to demonstrate that an integrated conservation-oriented programme which focused on conserving the natural resources of the Area could bring sustainable social and economic development to the local people and develop tourism with minimal adverse environmental impacts (Gurung, 1995). It owed its genesis to the perception that increases in population and trekking had led to degeneration of pastures and deforestation, erosion and landslides (the Himalayan environmental degradation theory), litter and human waste pollution, aberrations in local cultural values, and socio-economic inequalities. For the first 2 years the Project developed only slowly as the Ghandruk community reacted to a perception that they would lose control over their resources and find themselves directed by an outside agency as to what they could and could not do. Gradually, however, as the KMTNC gained the trust of the local people and the evidence of trekking fees and other assistance being returned back to the local community by way of development activities and conservation became obvious, their active participation ensued (Ghurmi, 1997).

Within 4 years, the Ghandruk pilot project had expanded to encompass 16 VDCs covering an area of more than 1500 km^2. Fees for trekking were introduced and in 1989 the Government enacted regulations to support KMTNC's management regime and to provide it with a guaranteed source of income by passing the trekking fees it collected to the Project. Two years later the boundaries of the ACA were enlarged to its present size, officially gazetted in 1992, and the KMTNC was granted responsibility to manage it for the next 10 years. By 1997, KMTNC had established seven field bases to work with the 55 village districts and some 122 Conservation Area Management Committees had been set up at the ward level within each of the Area's 55 VDCs to promote community development programmes, conserve the cultural and natural environment and to manage natural resources on a sustained-yield basis (Bajracharaya, 1996).

Three inter-related objectives guided the Ghandruk pilot project:

> (a) mitigating negative or undesirable environmental impacts through promotion of local guardianship and making tourism and other developmental activities responsive to the fragility of the area;
> (b) generating and retaining tourism and other sources of income in the local economy through skills development, increase in local production and local entrepreneurship; and
> (c) promoting linkages between conservation, tourism and local development through a pro-active approach to planning; ploughing back tourism revenue for local development, nature conservation and tourism development, and diversification of tourism products.
>
> (Adikhari and Lama, 1996: 3)

Recognizing that the conservation and protection of fragile habitats required the support of the local communities from the beginning, the ACAP concentrated on promoting 'peoples participation at the grass-roots level' in its activities (Adikhari and Lama, 1996: 2). This was a departure from then accepted practice in Nepal which, prior to 1986, had tended to follow the American management system which concentrated on the protection of wildlife and biodiversity with little or no regard to the welfare and role of local people. ACAP was the first major initiative in protected area management in Nepal where the capabilities of local people to contribute to conservation were recognized and attempted to be integrated into a conservation regime.

Inevitably in the introductory phases of the pilot project an ad hoc prescriptive approach tended to dominate the consultative and decision-making processes which were instigated: both ACAP staff and village representatives were on a learning curve as they grappled with the intricacies of an integrated, holistic model rather than a narrow sectoral approach (Yonzon, 1997). There is now more of a bottom-up approach, with the highly trained ACAP staff tending towards adjunct experts-oriented advice being fed into the decision-making process rather than dominating it. The current 5-year management plan places its emphasis on building human resources and strengthening institutional capacity at the local level, the ultimate objective being to devolve responsibility onto the local communities and empower them to undertake conservation activities on their own initiative. The five-year plan is based on three guiding principles:

1. Peoples' participation. Local communities are involved in planning, decision-making and implementation, with responsibility for managing local conservation areas. ACAP has formed various local user groups such as Conservation Area Management Committees, Womens' Committees, Lodge Management Committees, Forestry User Groups and so forth.

2. Catalysts or 'match-makers'. ACAP acts as a *lami* (match-maker) – the facilitator. It acts as a bridge between various international and national agencies and local communities to match the latter's needs to expertise and resources from the former.

3. Sustainability. Since emphasis is given to sustainability, those projects which can be implemented by local communities and maintained by them after external support is withdrawn are accorded priority. In the context of sustainability, local communities are expected to contribute 50% of the cost if possible and must demonstrate a commitment to the future management of the scheme.

To meet the goal of conservation for development, ACAP established four core programmes:

1. Resources Conservation Programme: The sustainable use and management of local natural resources, especially forests. Major activities include the establishment of the Conservation Area Management Committees, private and community nurseries, training in conservation, promotion of sound wildlife and soil and water management schemes, and the introduction of appropriate alternative energy programmes.

2. Sustainable Tourism Management: Formation of Lodge Management Committees, provision of information for tourists through publications and information centres, conservation of the environment especially forests, conservation of local culture and important heritage sites.

3. Sustainable Rural Development: Improving the basic living standard of local inhabitants mainly through agricultural development, infrastructure improvement, health and sanitation, women's empowerment programmes, youth programmes and programmes for the socially and economically deprived.

4. Conservation Education and Extension Programme (CEEP). This programme includes generation of awareness of conservation through education – accessing classes at local schools, running mobile awareness camps, publication and extension packages, networking with line agencies, public campaigns, study tours and home visits. This programme underpins the other three and is considered the backbone for the success of the ACAP.

In terms of forests, the Conservation Area Management Committees have been designated as the local institutions for making decisions about the use and management of existing forests. ACAP has stimulated these committees to revitalize community management systems of forests and other natural resources, and provided regular training for *ban herolas* (forest guards). In addition to encouraging local institutions to set up nurseries, plant trees and conserve forests, ACAP has distributed almost one million saplings, as mentioned above. In some districts the result has been dramatic – as the reforestation and plantations around Ghandruk demonstrate.

Increasingly forests are seen not only as sources of physical resources necessary to support tourists' needs and to meet local needs, but as places for recreation. The rhododendron forests around Sikles, where the ACAP has launched an ecotourism project, are a prime example of this and increasingly villages are constructing amenities (trails, resting sites, shelters) within forests for use by trekkers interested in wildlife. Sirubari, where the community established the first 'cultural immersion' village home-stay in Nepal in 1998 (a 3-day experience where visitors are encouraged to participate in the daily life of the Ghorka host families), has expanded its common forest by almost 100 ha, constructed paths to scenic lookouts, set up a rest shelter in the forest and meditation points for its visitors. Where in the past forest reserves had been set aside for the recreational pursuits of royalty and the ruling elite, under the impact of tourism community forests and protected areas are now becoming recreational resources for local residents and tourists. The KMTNC and the ACAP have led the way in this regard for Himalayan Nepal.

Parks and People Programme (PPP)

In the past decade it has become more widely accepted in Nepal that the dynamics of social structures and social processes constitute the context within which community forestry and protected areas management takes place. The Parks and People Programme, a partnership established in 1994 under an agreement between the United Nations Development Programme and the Government of Nepal (specifically the Department of National Parks and Wildlife Conservation), has attempted to address the problems of conflict and competing demands upon the resources of protected areas through the active participation of people in effective and sustainable park management and improvement in their socio-economic conditions (social mobilization). To some extent it has modelled its approach upon the ACAP. Activities of the PPP include reafforestation, skills enhancement training, income generation, micro-credit and micro-enterprises development, physical infrastructure improvement, conservation education and awareness programmes, park management, and demonstration and dissemination of appropriate rural technologies. Integral to the operations of the PPP has been the introduction in Nepal of buffer zones around protected areas, and the buffer zone around the Royal Chitwan National Park forms the focus of this examination of the development of community forests as one of the major activities of the PPP.

Amendments in 1996 to the 1973 National Parks and Wildlife Act provide the legal instrument to establish and define buffer zones in Nepal as the area surrounding a park or reserve encompassing forests, agricultural lands, settlements, village open spaces and many other land use forms. Management is to focus on compatible land use and sustainable development to meet the special needs of communities living within the Zone. A key aim is to find ways for the protected areas to contribute to the well-being of the buffer zone communities, thus creating a climate where the conservation objectives of the protected areas will be supported rather than opposed and/or violated. This emphasis on socio-economic development is a rather different concept from the traditional understanding of a buffer zone where the strategy is to create low-use areas surrounding the protected area. To give some efficacy to the buffer zone concept the Act provides for 30–50% of all revenues generated by the park or reserve to be retained for community development.

Nepal's Buffer Zone Management Regulations and Development Guidelines provide mechanisms to mobilize villagers' participation in community development:

- the households in a distinct settlement (called 'a unit') are to be mobilized to form a 'User Group' (UG);
- a User Group or several user groups are to form a User Committee (UC) with a minimum of nine members drawn from the settlement unit;
- the User Committees are to perform coordinating and supporting roles between User Groups and the Buffer Zone Office (headed by the buffer zone Warden) of the Department of National Parks and Wildlife Conservation (DNPWC) to mobilize resources and implement programmes;
- the User Committees are to facilitate the flow of the share of government revenue committed for community development to fund proposals submitted by User Groups;
- the buffer zone Warden is the point of official contact for various UC offices spread throughout the buffer zone;
- there is to be a Buffer Zone Development Council composed of all of the chairpersons of the UCs, with the buffer zone Warden as the ex-officio member secretary.

The role of the buffer zone Warden is seen as crucial in facilitating the formation of the UGs/UCs, seeking cooperation from line ministries to undertake a range of activities in the buffer zone and actively working with INGOs such as the UNDP to promote a growth pattern consistent with the protected area's primary objectives of conservation.

In terms of the Royal Chitwan National Park, the buffer zone encompasses about 750 km^2. The Zone includes 34 Village Development Committees and 128 village wards, representing a total population of about 250,000. In 1997, the Department of National Parks and Wildlife Conservation (DNPWC) passed about R22 million to the PPP for distribution to UCs for projects being implemented by UGs, of which there are now more than 250 (Parks and People Programme, Annual Report 1997). These funds are derived from royalties received from seven concessions granted to private enterprises to operate wildlife lodges within the Park, entrance fees from visitors, and fees from the hire for safaris of DNPWC's 22 elephants based at its Sauraha headquarters in the Chitwan Buffer Zone.

The PPP has also formulated an ecotourism management plan for the Chitwan Buffer Zone, much of it based around community forests. A major activity of the PPP is support for reafforestation and the establishment of community forests in the Buffer Zone, in order to reduce dependency upon the forest resources of the Chitwan National Park. It works with Forest User Groups (FUGs) which set up committees elected by the community concerned. The model adopted for the Chitwan buffer zone has committee members (with a minimum of three women) elected for a 5-year term by the FUG (which itself consists of at least one member from each household within each ward covered by the VDC). The FUG committee is able to be dismissed if a majority of the user group considers there is just cause (such as mismanagement, corruption, incompetence). The committee is given responsibility to produce an annual programme which is presented to the FUG as a whole and implemented only after consensus acceptance. It meets regularly, generally once each month, to monitor implementation of its programme and to take decisions regarding access by community members to the forest, sale of forest products, and penalties for any transgressors of its local regulations. It must maintain records of commercial transactions on behalf of the FUG as a whole and it designs development proposals for any funds raised. The active participation of community members is thus assured in the management of their forests.

By 1997, the PPP, working with Forest User Groups, had developed 555 ha of community plantations within the Chitwan buffer zone, with more developed by other NGOs such as the KMTNC. The latter, for example, has worked for the past 10 years with the Bacchauli Village Development Committee and its Forest User Group to rehabilitate the highly degraded, over-grazed and over-logged 400-ha Baghmara Community Forest, located in the north-eastern boundary of Chitwan. During 1997, under the PPP, more than 550,000 saplings were distributed to Forest User Groups which planted them (with an 85% survival rate) on 250 ha of community lands, private land, along roadside verges, and around public areas such as school grounds (PPP, 1998). In addition a 60-ha forest (Nirmalbasti Community Forest) was handed back to community control at Parsa.

These community forests have relieved very considerable pressure on the forest resources of Chitwan. From Baghmara Community Forest alone, for example, it was estimated that in 1992 over 90,000 kg of grass were harvested for thatching, more than 225 tonnes of fuelwood was provided from the plantation area of the forest (an estimated 50% of community needs), and more than 2.4 million kg of fodder was collected (Rijal, 1997). All of Bacchauli's fuelwood needs are expected to be met by Baghmara Forest as regeneration increases density over the next 5 years.

As numbers of animals and birds increase inside the Park under effective National Parks/Army protection and spill over into the surrounding farmlands, the community forests provide adjunct habitats and sanctuary. For example, in 1960 there were only 60 rhinos remaining in Chitwan. That number had increased to 466 by 1995, and about eight to ten rhinos have migrated out of Chitwan and are now living permanently in Baghmara. A 1975 survey indicated that only about 40 tigers remained in Chitwan; by 1995 the number had increased to more than 120. The sizeable population of prey species such as sambur deer, barking deer, hog deer, spotted deer, rhesus macaques, langurs, wild boar and others now inhabiting Baghmara and other community forests also provide sustenance for carnivores such as the tiger and leopard, both of which have been recorded in community forests (Rijal, 1997). More than 200 bird species have also been recorded in community forests of the Chitwan Buffer Zone. These forests are proving an increasingly useful natural absorbent which serves to decrease conflict as animals leave the Park in search of territories and food.

The PPP has encouraged ecotourism to the Community Forests as a means by which the local people can generate income. In and around Sauraha buffer zone there are now more privately owned elephants than the 22 in the service of DNPWC for elephant safaris into Chitwan. The privately owned elephants are not permitted to cross the Rapti River into the Park (the seven private concessions, together with the DPNWC, have exclusive rights to provide elephant safaris inside the boundaries) and so they utilize community forests with the permission of the Forest User Groups. An elephant which costs about US$15,000 to purchase currently earns about the same per year for Sauraha owners so over a period of years the return is significant (Sofield and Bhandari, 1998). A number of small lodges have been built adjacent to community forests in the buffer zone, and several Forest User Groups are currently exploring the idea of constructing *machans* (viewing towers with two rooms and four beds) inside the community forests for ecotourism. More than 120 guides have been trained in nature/environmental interpretation and guiding under short courses run by the PPP. Other PPP-sponsored projects for ecotourism include skill enhancement courses (particularly for women) to produce handicrafts from forest products, bamboo craft, Dhaka weaving and cultural activities. In this context it has established a Womens User Group handicrafts retail outlet in the grounds of the DNPWC headquarters at Sauraha in association with the International Union for the Conservation of Nature (IUCN). Revenue and employment generated by the tourism industry has heightened awareness and understanding among local residents of the benefits of management and conservation of their forest resources, and thus decreased the incidence of illegal incursions, poaching and logging in Chitwan National Park. With support from institutions such as the PPP and KMTNC the community forests of the Chitwan buffer zone provide a model which is economically and ecologically sustainable. Equally important is the social mobilization of the communities themselves and their empowerment through effective local control over their forest resources.

Conclusions

The forests of Nepal have been an important resource for recreation for centuries, and the modern phenomenon of tourism is now adding to that traditional role. Because of the global growth in ecotourism the forests of Nepal are being utilized in a novel way for economic development by schemes which encourage the participation of rural villages in this form of tourism. The regime of national parks is also focused not only on conservation and maintenance of biodiversity but on nature-based tourism. As well as providing essential materials and products for the survival of its rural communities, Nepal's forests are thus now playing an important new role in the development process. As Gilmour and Fisher (1992: 78) noted:

> Attempts to intervene in resource management always result in social processes developing with the community. It is only by explicitly recognising these processes and matching them with attempts to assist with consensus building and institution building that community forestry can claim to be a special form of people centred forestry.

The models of community participation in forestry management and ecotourism developed by the King Mahendra Trust for Nature Conservation in the Annapurna Conservation Area, and of the Parks and People Programme in the Chitwan Buffer Zone, meet these basic requirements and therefore have the potential to be sustainable in economic, ecological, socio-cultural and touristic contexts. The role of tourism in underpinning these efforts by providing a sound economic foundation which can penetrate local communities is essential to the sustainability of conservation and protection of the forests and biodiversity of Nepal.

References and Further Reading

ACAP (1998) *Annapurna Conservation Area Project Annual Report.* ACAP, Nepal.

Adikhari, J. and Lama, Tshering T. (1996) *Annapurna Conservation Area Project. A New Approach in Protected Area Management. A Decade of Conservation for Development (1986–1996).* King Mahendra Trust for Nature Conservation (KMTNC), Kathmandu.

Akhtar, S. and Karki, A.S. (eds) (1998) *Issues in Mountain Development,* 98/3. International Centre for Integrated Mountain Development (ICIMOD), Kathmandu.

Allen, Charles (1999) Nepal Chronicle. In: Cheogyal, L. and Moran, K. (eds) *Nepal.* APA Publications, Basingstoke, pp. 61–68.

Bajracharya, S.B. (1983) Fuel, food or forests? Dilemmas in a Nepal village. *World Development* 11, 1057–1074.

Bajracharya, S.B. (1996) An overview of Annapurna Conservation Area Project. In: *Proceedings of the 10th Anniversary of KMTNC.* KMTNC, Kathmandu.

Baker, G. (1995) *Nepal – Sagarmatha Forestry Report 1995 Visit.* New Zealand Forest Research Institute, Rangiora.

Banskota, K. and Sharma, B. (1995a) Mountain tourism in Nepal: an overview. Discussion Paper Series No. MEI 95/7. ICIMOD, Kathmandu.

Banskota, K. and Sharma, B. (1995b) Economic and natural resource conditions in the districts of Baghmati Zone and their implications for the environment: an adaptive simulation model. Discussion Paper Series No. MEI 95/9. ICIMOD, Kathmandu.

Banskota, K. and Sharma, B. (1995c) Tourism for mountain community: case study report on the Annapurna and Gorkha Regions of Nepal. Discussion Paper Series No. MEI 95/11. ICIMOD, Kathmandu.

Banskota, K. and Sharma, B. (1995d) Carrying capacity of Himalayan resources for mountain tourism development. Discussion Paper Series No. MEI 95/14. Kathmandu: ICIMOD, Kathmandu.

Banskota, K. and Sharma, B. (1998) Mountain tourism for local community development in Nepal: a case study of Upper Mustang. Discussion Paper Series No. MEI 98/1. ICIMOD, Kathmandu.

Banskota, K. and Sharma, B. (1996) *Royal Chitwan National Park: An Assessment of Values, Threats and Opportunities.* KMTNC, Kathmandu.

Bauer, Johannes (1995) *Pharak Community Forestry Project.* World Wildlife Fund (WWF), Kathmandu.

Bornemeier, J., Victor, M. and Durst, P. (eds) (1997) *Ecotourism for Forest Conservation and Community Development.* Proceedings of an International Seminar, Chiang Mai, Thailand, 28–31 January, 1997. FAO/RAP Publications, Bangkok.

Bromley, D.W. (1986) On common property regimes in international development. Paper presented at the AKRSP/ICIMOD/EAPI Workshop on Institutional Development for Local Management of Rural Resources, Gilgit, Pakistan. ICIMOD, Kathmandu.

Bruijnzeel, L.A. (1989) Highland-lowland Interactions in the Ganges Brahmaputra River Basin: A Review of Published Literature. ICIMOD Occasional Paper no. 11, ICIMOD, Kathmandu.

Campbell, J.G., Shrestha, R.J. and Euphrat, F. (1987) Socio-economic factors in traditional forest use and management. Preliminary results from a study of community forest management in Nepal. *Banko Janakari* 1, 45–54.

Carter, A.S. and Gilmour, D.A. (1989) Tree cover increases on private farm land in central Nepal. *Mountain Research and Development* 9, 381–391.

Department of National Parks and Wildlife Conservation (1998) Annual Report 1997. DNPWC, Kathmandu.

Dove, M.R. and Rao, A.L. (1986) Common resource management in Pakistan. Paper prepared for AKRSP/ICIMOD/EAPI Workshop on Institutional Development for Local Management of Resources, 18–25 April, Gilgit, Pakistan.

East, P., Luger, K. and Inmann, K. (eds) (1998) *Sustainability in Mountain Tourism: Perspectives for the Himalayan Countries.* Book Faith India, New Delhi.

Eckholm, E.P. (1975) The deterioration of mountain environments. *Science* 189, 764–770.

Fisher, R.J. (1990) The Himalayan dilemma: finding the human face. *Pacific Viewpoint* 31, 69–76.

Fisher, R.J., Singh, H.B., Pandey, D.R. and Lang, H. (1989) The management of forest resources in rural development – a case study of Sindhu Palchok and Kabhre Palanchok Districts of Nepal. Mountain Populations and Institutions, Discussion Paper no. 1. ICIMOD, Kathmandu.

Furer-Haimendorf, C. von (1964) *The Sherpas of Nepal: Buddhist Highlanders.* University of California Press, Los Angeles.

Furer-Haimendorf, C. von (1984) *The Sherpas Transformed: Social Change in a Buddhist Society of Nepal.* Sterling Publishers, New Delhi.

Gautam, K.H. (1987) Legal authority of forest user groups. *Banko Janakari* 1, 1–4.
Ghurmi, G. (1997) Developing a tourist destination: the experience of the King Mahendra Trust for Nature Conservation with ecotourism. In: Bornemeier, J., Victor, M. and Durst, P. (eds) (1997) *Ecotourism for Forest Conservation and Community Development.* Proceedings of an International Seminar, Chiang Mai, Thailand, 28–31 January, 1997. FAO/RAP Publications, Bangkok, pp. 176–186.
Gilmour, D.A. (1988) Not seeing the trees for the forest: a re-appraisal of the deforestation crisis in two hill districts of Nepal. *Mountain Research and Development* 8, 343–350.
Gilmour, D.A. and Fisher, R.J. (1992) *Villagers, Forests and Foresters. The Philosophy, Process and Practice of Community Forestry in Nepal.* Sahayogi Press, Kathmandu.
Goldstein, M.C., Ross, J.L. and Schular, S. (1983) From a mountain-rural to a plains-urban society: implications of the 1981 Nepalese census. *Mountain Research and Development* 3, 61–64.
Griffin, D.M. (1987) Intensified forestry in mountain regions. *Mountain Research and Development* 7, 254–255.
Gurung, C. (1995) People and their participation: new approaches to resolving conflicts and promoting cooperation. In: McNeely, J.A. (ed.) *Expanding Partnership in Conservation.* Island Press, New York.
Hardin, G. (1968) The tragedy of the commons. *Science* 162, 1243–1248.
HMGN (1986) Land use in Nepal – a summary of the Land Resources Mapping Project Results. Report No. 4/1/310386/1/1 Seq. No. 225. Ministry of Water Resources, Water and Energy Commission, Kathmandu.
Hobley, M. (1985) Common property does not cause deforestation. *Journal of Forestry* 83, 663–664.
Ingles, A. (1990) The management of religious forests in Nepal. Graduate Diploma thesis, Australian National University, Canberra.
Ives, J.D. and Messerli, B. (1989) *The Himalayan Dilemma – Reconciling Development and Conservation.* The United Nations University and Routledge, London and New York.
King Mahendra Trust for Nature Conservation (KMTNC) (1998) *Technical Report on Buffer Zone Policy Analysis of the Royal Chitwan National Park.* KMTNC, Lalitpur.
Lama, Wendy Brewer and Sherpa, Ang Rita (1995) Tourism Management Plan for the Upper Barun Valley. The Makalu-Barun Conservation Project Working Paper Publication Series Report 24. Department of National Parks and Wildlife Conservation, Kathmandu.
Ledgard, N.J. (1997) Nepal – Sagarmatha Forestry Report 1997 Visit. New Zealand Forest Research Institute, Rangiora.
Land Resources Mapping Project (1986) *Land Utilization Report. Land Resources Mapping Project.* HMGN, Kathmandu.
Mahat, T.B.S., Griffin, D.M. and Shepherd, K.R. (1986a) Human impact on some forests of the Middle Hills of Nepal. 1. Some major human impacts before 1950 on the forests of Sindhu Palchok and Kabhre Palanchok. *Mountain Research and Development* 6, 223–232.
Mahat, T.B.S., Griffin, D.M. and Shepherd, K.R. (1986b) Human impact on some forests of the Middle Hills of Nepal. 2. Forestry in the context of traditional resources of the state. *Mountain Research and Development* 6, 325–334.

Messerschmidt, D.A. (1987) Conservation and society in Nepal: Traditional forest management and innovative development. In: Little, P.D., Horowitz, M.M. and Nyerges, A.E. (eds) *Lands At Risk in the Third World: Local Level Perspectives.* Westview Press, Boulder, Colorado, pp. 373–397.

Ministry of Tourism & Civil Aviation (1998) Annual Report, 1998. HMGN, Kathmandu.

Mishra, H.R. and Jefferies, M. (1991) *Royal Chitwan National Park: Wildlife Heritage of Nepal.* The Mountaineers, Seattle.

Molnar, A. (1981) The dynamics of traditional systems of forest management, Nepal. Report to the World Bank (unpublished).

National Planning Commission (1998) Ninth Development Plan. HMGN, Kathmandu.

Netherlands Development Organization (SNV) (1998) SNV-Nepal Annual Plan 1999. SNV, Kathmandu.

Parks & People Programme (1998) Annual Report 1997. HMGN/UNDP, Kathmandu.

Ramble, C. (1999) Natural crossroads. In: Cheogyal, L. and Moran, K. (eds) *Nepal.* APA Publications, Basingstoke, pp. 37–44.

Regmi, M. (1978) *Land Tenure and Taxation in Nepal.* Bibliotheca Himalayica Series 1. Ratna Pustak Bhandar, Kathmandu.

Rieger, H.C. (1981) Man versus mountain: the destruction of the Himalayan ecosystem. In: Lall, J.S. and Moddie, A.D. (eds) *The Himalaya: Aspects of Change.* Oxford University Press, New Delhi.

Rijal, Arun (1997) The Baghmara Community Forest: an example of linkages between community forestry and ecotourism. In: Bornemeier, J., Victor, M. and Durst, P. (eds) (1997) *Ecotourism for Forest Conservation and Community Development.* Proceedings of an International Seminar, Chiang Mai, Thailand, 28–31 January, 1997. FAO/RAP Publications, Bangkok, pp. 144–150.

Rogers, P. and Aitchison, J. (1998) *Towards Sustainable Tourism in the Everest Region of Nepal.* IUCN Nepal, Kathmandu.

Shah, D. and Shrestha, T.B. (eds) (1996) *Biodiversity Databases of Nepal.* IUCN, Kathmandu.

Sofield, T.H.B. (1992) The Guadalcanal Rainforest Trail: ecotourism in Melanesia. In: *Ecotourism Business in the Pacific: Promoting a Sustainable Experience.* Auckland University and East West Center, University of Hawaii, Auckland, pp. 89–100.

Sofield, T.H.B. and Bhandari, S. (1998) *An Independent Evaluation of the Partnership for Quality Tourism Programme, UNDP, Nepal.* UNDP, Kathmandu.

The Mountain Institute (1998) Annual Report 1997. TMI, Franklin, USA.

Thompson, M., Warburton, M. and Hatley, T. (1986) *Uncertainty on a Himalayan Scale.* Milton Ash Editors, Ethnographica, London.

UNDP (1994) Parks and People Project Document. HMGN and UNDP, Kathmandu.

World Bank (1978) *Nepal Staff Project Report and Appraisal of the Community Forestry Development and Training Project.* World Bank, Washington, DC.

Yonzon, Pralad (1997) Ground-truthing in the protected areas of Nepal. In: Bornemeier, J., Victor, M. and Durst, P. (eds) (1997) *Ecotourism for Forest Conservation and Community Development.* Proceedings of an International Seminar, Chiang Mai, Thailand, 28–31 January, 1997. FAO/RAP Publications, Bangkok, pp. 82–94.

15

Planning for the Compatibility of Recreation and Forestry: Recent Developments in Woodland Management Planning within the National Trust

David Russell

Introduction

When John Evelyn consolidated the principles of silviculture in the 17th century, forestry was already concerned with the management of woods and plantations for a variety of private and public benefits. Really only in the immediate post-war period of this century did forestry acquire a narrow focus on timber production, and that has often been more in the theory than the practice.

I worked for some years in commercial forestry in southern England. Most resident woodland owners enjoyed field sports more than timber production. Many absentee owners were more than happy to allow public access to their plantations and during the hot dry summer of 1976 several potentially devastating forest fires were caught early owing to the vigilance of the public. So there is little justification for tension between public recreation and forestry.

What does matter is the tensions that potentially exist between private rights and public interest (whose countryside is it?), between competing recreational interests, and between recreation and the interests of conservation.

I want to offer a general account of how we try to establish compatibility between all the interests in what is a genuinely multipurpose woodland enterprise in the National Trust. I will describe our general philosophy of woodland and recreation management and then spend some time describing the planning process which we are currently developing, as an environmental management system. The National Trust has 2.5 million members and owns 250,000 ha of land (1% of the UK). Its purposes are (The National Trust, 1971):

> to promote the permanent preservation for the benefit of the nation of land and tenements (including buildings) of beauty or historic interest and as regards lands for the preservation (so far as practicable) of their natural aspect features and animal and plant life.
>
> (1907 NT Act)

Part of the Trust's business is to

> Provide access to and enjoyment of places of natural interest and beauty.
>
> (1937 NT Act)

I mentioned tension between private rights and public interest, and you will see that the statutory purposes of the Trust already signal a singular compromise. The Trust owns land but is required to manage it for the benefit of the nation.

We have also decided on our position should there be any tension between recreation and conservation interests: as a result of our 1995 Access Review we have adopted the robust view that 'If serious conflict arises, conservation will take precedence over access'. Normally there is no problem, but sometimes there may be temporary closing of part of a wood, for example where a rare bird is nesting.

Forestry

We own and manage 27,000 ha of woodland in England, Wales and Northern Ireland. Unlike other large forestry interests (Forestry Commission, Woodland Trust, etc.), forestry is a small part of our portfolio of land management and our forests are not managed as a separate entity from the rest of our land. There is no separate forestry department, and relatively few staff totally dedicated to forestry work.

Broadly speaking our forestry objectives are:

- conservation of woods as features in a landscape,
- protection of distinctive aesthetic characteristics of woods (or spirit of place),
- conservation of wildlife (or biodiversity),
- protection of historic features and
- provision of some form of recreational opportunity and usually (but not always) unrestricted public access without charge.

Timber production is not a priority but nevertheless generates an income of around £500,000 per annum and we aim to sustain our capacity to produce timber.

Recreation

We manage for public recreation as part of our core purpose. However, unlike other access providers we do not, on the whole, market recreational packages. Our speciality is tranquillity. Our brand image is quiet enjoyment of the countryside. We make places available, provide car parks where appropriate and maintain paths free from hazards, so far as possible.

Signing and interpretation panels are usually sited near car parks, but in keeping with our brand image we prefer them to be discrete and unassuming. Our aim is to intrude as little as possible in people's experience of the place.

However, I do not want to give the impression that we are opposed to more active types of enjoyment. We are, in principle, open to almost any legal activity so long as it can be managed (within reason) so as not to interfere with the enjoyment of others.

In 1995 we published 'Open Countryside'; the report of a review of access and recreational policy and practice in the National Trust. We were attempting to discover from user groups and others how well we provided recreational opportunity and whether we could do better.

The review also assessed the impact of recreation on the effectiveness of conservation of special sites or habitats. The view was that any negative impact was relatively localized. Woods, as you would expect, are amongst the most resilient of sites able to absorb many people and many different recreational activities without compromising conservation interests.

The review has helped to resolve potential conflict between different recreational interests, firstly by establishing regular channels of communication with representative bodies and secondly by encouraging them to promote their codes of conduct more effectively.

There are two more general points that I want to make about our approach to recreation. Firstly we are concerned about what might be called, 'barriers' to access. These might be economic, cultural or infra-structural. Perhaps people cannot afford to get to places where recreation is possible, perhaps they do not know how to or feel threatened in some way if they do and perhaps they simply lack the means. We do not want to force people to visit us, but we do not want them to be prevented if they want to come.

We have trialed some inner city projects to help with this. One has been in Newcastle. We have engaged with kids who have never played football on grass. We have a lot of grass and a lot of woods. When they came they persuaded us that an assault course would make the woods more interesting, so we helped them build one. Older people, too, are involved. We have helped some of them set up a group called the Walker walkers. (Walker is a district of Newcastle.) From a starting point of insecurity in the countryside they have grown in confidence and now travel all over the north of England for walking trips. We lease them a minibus but otherwise they are now on their own. We have very few such schemes and this scarcely begins to touch the tip of an enormous urban iceberg.

The second point is that we have corporately adopted a set of environmental principles:

- to protect our long-term interests from environmental damage;
- to promote the wider protection of the environment;
- to avoid creating adverse environmental impact through what we do.

The present heavy reliance on public transport to get to places for recreation presents us with a dilemma. We want people to come, and houses and parks rely upon the income we derive from ticket sales, membership and enterprise outlets of various types, but we recognize that the use of the car is contributing to physical damage to buildings, which is very expensive to remedy, and physical damage to the countryside, which is probably impossible to remedy.

Public transport is not good enough to replace the car, though we have been actively promoting and supporting a number of integrated transport programmes. We promote the bicycle and even provide them at some holiday cottages. But increasingly we see the need to, as it were, take the opportunities for access to the town. There is need for more urban green space, despite the emphasis on development of brown-field sites for new houses. And we also need to consider whether some of our future acquisitions should be consciously urban with a view to providing particularly for access.

Recreation in Woods

While there is little or no conflict between recreation and forestry there are certainly demands on resources. Occupiers of land are liable for the safety of people on their land. This means we have to undertake regular risk assessments.

Some years ago we prepared guidance on tree inspection procedures. Case law has established that National Trust staff while not expected to be expert arboriculturists, should nevertheless be more skilled than the average countryman. Key staff managing access in wooded areas are trained in basic arboriculture and risk assessment techniques and have to undertake annual inspections of all trees in high risk areas.

Almost all our woodlands and parks are open to the public with usually unrestricted access. Some recreational activities are closely monitored. For example, at one large woodland in Hertfordshire, horse riding is by permit only and the Trust has employed one of the local riders as a warden.

Occasionally, we permit and even sometimes charge for certain recreational activities which are far from tranquil or demand special facilities, like car or motorcycle rallies. We also let shooting rights in some woodlands (though it is difficult to argue that this is much to the benefit of the nation). Orienteering is a regular and popular activity. Mountain biking is ubiquitous and one of the most unpopular activities with walkers. Most recreation is informal but periodically most property staff arrange a guided walk.

The major exception to our non-packaging of woodland recreational opportunity is our 'working holiday' programme for volunteers. The Trust works with over 38,000 volunteers each year and woodland work is popular. Under this scheme we put together a varied programme of work within easy reach of a base camp where the volunteers stay. The quality of accommodation is good. They tackle rhododendron clearance, scrub bashing, thinning of young trees and respacing or natural regeneration and sometimes tree planting. They have a lot of fun and so do we. The Trust's access and recreational management work in woods is supported very considerably by Forestry Commission grant aid.

Woodland Management Planning

Let me try to put this into the context of a programme of woodland management planning, for it is here that any tensions between different recreational interests will become evident. It is here, too, if things are done well, that the tensions can be resolved, almost always by some sort of physical or temporal zoning with the agreement of interested individuals or representative groups. A key component of our developing management planning programme is dialogue with and between all the main stakeholder groups. We have adopted a general approach to property management planning and try to make the connections to forestry and recreation wherever necessary. In the management of woods we start with a framework based on a set of questions:

What is this place/how is it used?	Significance
What happens if we do nothing?	Process
What could it be?	Vision
What must we do?	Action
Did it work?	Audit

The elements can be set out in a cyclical form (Fig. 15.1). I want to take each point in the circle and elaborate.

Significance. Conservation has been described (Holland and Rawles, 1994) as being about 'negotiating the transition from past to future in such a way as to secure the transfer of maximum significance'. Conservation is about what matters.

The starting point is stakeholder dialogue, consultation and specialist survey. We need to know what we have, how it is used and what matters to us and to others and what is the potential for enhancing its value or significance. This will then be summarized in a statement of significance.

The process is important as the basis for an ongoing dialogue with communities of interest. The process should be conducted with an open mind and a willingness to reconsider established views.

Most importantly we have to look outside the boundaries of the property. This is something that most forestry management plans overlook. Boundaries are irrelevant when searching for significance. Significance does not respect boundaries.

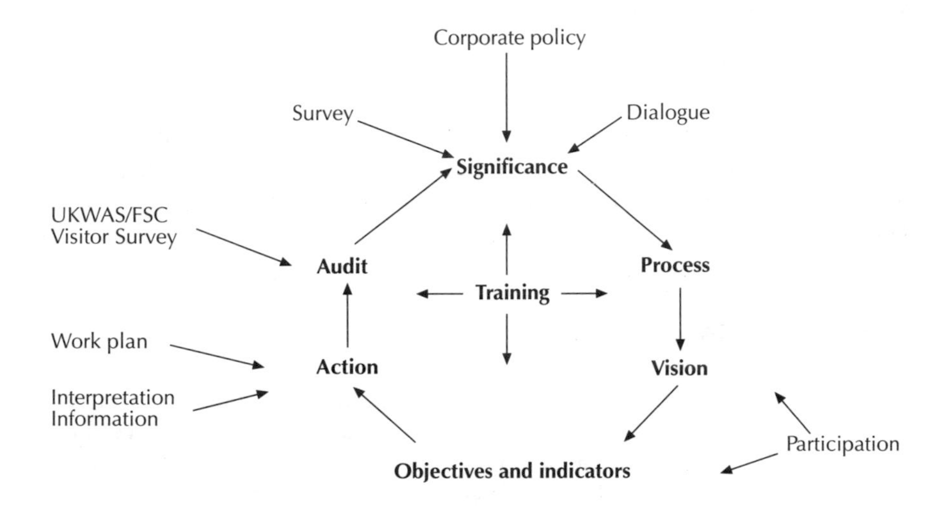

Fig. 15.1. Woodland management planning cycle.

The statement must be sufficiently well done to serve a number of purposes:

- provide, so far as practicable, a summary of all the most significant features and attributes (including uses);
- provide a fair recognition of the range of opinion about the significance of the place;
- provide a sufficient basis for all property management planning, especially objective setting;
- help members of the public or the relevant agencies who may not appreciate the range of features and attributes at a property or may not agree with subsequent management proposals;
- be succinct, clearly presented and inspiring;
- be accessible and meaningful to a non-specialist readership;
- be widely disseminated.

These sorts of things might be included in the summary of features and attributes:

Sense of place	Whatever is distinctive and evocative about a property and its environs. Symbolic importance
Historical or documentary	Features (artefacts and natural) which individually and collectively record the history of the place and the people who have lived and worked there

Human health and environment	Features and activities which are critical elements of a healthy environment; including impacts on the property and those originating at the property
Biodiversity	Habitat and species characteristic of the place and the factors which influence their survival, past biodiversity and the geological and fossil record
Educational	Features or activities which provide educational opportunities
Social	Features or activities which provide a service to local people; features which have special significance for local people; activities which provide opportunities for public participation
Economic	Features or activities which generate revenues for the Trust; features or activities which generate revenues for local business Activities which reduce resource demands (energy, water, etc.) and promote local production
Recreational	Features which provide opportunity for people to do things individually and activities which promote fun, delight and inspiration

I would not wish to claim that all these factors are considered in every case, the process is still developing. However it is on the basis of such an analysis that we must decide what really matters.

Process. This is asking the question, what happens if we do nothing? It is designed to counter a tendency we all have to over-manage. An anxiety to tidy up woodlands in south-east England after the storm of 1987 led to more damage and a lot of unnecessary tree planting. So the question is designed to encourage people to understand the natural processes which are shaping the woodland ecosystem. Let us work with nature rather than fighting it. It has an economic as well as an ecological rationale.

Vision. Without a sense of what is desirable it would be impossible to set objectives. We have to take the broadest possible view, intellectually and physically, because much will depend on what is going on around us. In the case of recreation this might well include a response to new pressures or opportunities.

Objectives and indicators. These follow from a consideration of significance and vision. Too often, not just in the Trust, objectives are set without a proper consideration of what the business of the enterprise is. The significance and vision processes are important in making the objective setting more transparent. Objectives, once agreed, need indicators to enable managers to measure progress.

Action. This includes both physical action on the ground and whatever programme of interpretation, education, community participation and continuing dialogue is appropriate.

In the case of woodlands I am keen on relatively little action on the ground in many cases, but it is difficult to persuade people that nature does not always need the intensive mollycoddling that we want to give it. Trees can and do regenerate without great planting jamborees. Too often our interventions are ill-judged and counterproductive.

Audit. This is easily overlooked but is a key part of planning, using the indicators set in the plan to assess the effectiveness of actions in meeting objectives over time. My job, and that of my advisory colleagues, has a major component of auditing and finding the best means to deal with the weaknesses exposed by audit. In our woodland management we will shortly be inviting external independent audit through the UK Woodland Assurance Scheme and FSC (Forest Stewardship Council).

Up to now infrequent visitor surveys on selected sites have provided our only feedback on visitor satisfaction. Constructive and regular dialogue with stakeholders, whatever the difficulty in setting it up and keeping it going, will need to be more informative.

Training is provided to support all the stages in the planning and implementation of forestry and recreation management. Current programmes (1998/9) include:

- community participation techniques
- managing volunteer programmes
- dog workshop
- horse workshop
- cycling workshop
- introduction to statements of significance
- community workshops
- management planning and objective setting
- 'Nature in transition' conference 1999
- countryside interpretation skills
- tree inspection and risk assessment.

Conclusions

We do not see any serious problem of compatibility between recreation and forestry. Problems, where they exist, are principally between private rights and public interest (not a problem for us), between some different recreational activities (resolved by regular dialogue, monitoring and zoning) and between recreation and conservation (relatively rarely a problem but resolved by temporary closure).

Management planning needs to be based on a 'Statement of Significance' summarizing all the ways in which the wood is used and valued. It is the essential basis for a continuing dialogue with stakeholders and for building compatibility between all the components in a genuinely multipurpose forestry enterprise.

References

Holland, A. and Rawles, K. (1993) Values in Conservation. *Ecos* 14, 14–19.

The National Trust (1971) The National Trust Acts 1907 to 1971 with Extracts from Certain Public Acts. The National Trust, London.

The National Trust (1995) Open Countryside, Report of the Access Review Working Party. The National Trust, London.

16

Beyond Carrying Capacity: Introducing a Model to Monitor and Manage Visitor Activity in Forests

Simon McArthur

Introduction

One of the principal reasons why we still have forests is because people have ascribed some form of value to them. These values come in part from vicarious appreciation and partly from direct use. Vicarious appreciation is driven by the belief that things have an intrinsic right to exist in perpetuity, and, by the notion that it is just 'nice to know it's there'. Value through direct use is driven by using the forest to derive benefit. For example, we keep many forests to provide clean air and water that we need for a healthy life.

Another form of direct use of forests is to personally visit and experience them. This experience can involve viewing, hearing, seeing, touching and even eating parts of a forest. The degree of direct benefit a person obtains from experiencing a forest is typically the result of the quality of their experience. A defining element of the quality of most forest-based experiences is the condition of the forest resource. But when people experience a forest, they are highly likely to change it in some way (Hammit, 1990; Glasson *et al.*, 1995). For example, they may trample vegetation, start a fire or disturb wildlife. This change often reduces the value of the forest, which, in turn, may lead to a decline in visitor satisfaction and the benefit that others can gain (Hall and McArthur, 1996). The behaviour of one visitor experiencing a forest can also influence the satisfaction and benefit of another experiencing the same forest. For example, they may interrupt a camp site or picnic area with drunken behaviour. Therefore, the condition of a forest and the quality of the visitor experience are inextricably linked to the way people go on valuing the forest and supporting its conservation (Hammit, 1990). A good deal of the way visitors are managed comes from the recognition of this relationship. In fact, the basic definition of visitor management as 'the management of visitors in a manner that maximizes the quality of the visitor experience while assisting the achievement of the area's overall management objectives' (Hall and McArthur, 1996: 303) reflects the need to keep this relationship in check.

Table 16.1. Qualitative assessment of visitor management techniques.

Visitor management techniques	Visitor management performance		Other aspects of performance		
	Conservation of forest	Improve quality of visitor experience	Create support for forest management	Proactiveness	Management reliance on technique
Regulating access	◆◆◆	◆	•	•	•
Regulating visitation	◆◆	◆◆	•	•	••
Regulating behaviour	◆◆	◆	•	•	•
Regulating equipment	◆◆	◆◆	•	••	••
Entry or user fees	◆◆	◆	•	••	•••
Modifying the site	◆◆	◆◆	••	•	•••
Market research	◆◆	◆◆	•	•••	•
Visitor monitoring and research	◆◆	◆◆◆	•	•••	•
Marketing – promotional	◆◆◆	◆◆	•••	•••	•
Marketing – strategic information	◆◆◆	◆◆◆	•••	••	••
Interpretation	◆◆◆	◆◆◆	•••	•••	••
Education	◆◆◆	◆◆	••	•••	•
Presence of forest managers	◆◆◆	◆◆	•••	••	•
Alternative providers – tourism industry	◆◆	◆◆◆	••	•••	•
Alternative providers – volunteers	◆◆◆	◆◆◆	•••	•••	••
Favoured treatment for accredited bodies bringing visitors to a site	◆◆◆	◆◆◆	•••	•••	•

Performance in relation to visitor management: ◆ limited performance; ◆◆ reasonable performance; ◆◆◆ good performance.
Performance in relation to other criteria: • limited; •• reasonable; ••• good.

This chapter therefore stresses that in order to move towards sustainable forest use, we need to improve and better integrate visitor management into the environmental management of forests.

Conventional Approaches to Managing Visitors

There are many visitor management approaches used at forest sites. Some of these approaches are focused on the site, some on the visitor and a few on other stakeholders. Most of the conventional approaches used by forest managers are based on regulating or controlling visitor access and behaviour, such as:

1. regulating access by area, time of year or day, transport;
2. regulating visitation through limitations on numbers, group size and type of visitor;
3. regulating behaviour through limitations on transport, equipment;
4. controlling the cost of visiting through entry or user fees; and
5. modifying the site through hardening the resource, building more facilities and erecting barriers.

These approaches are popular with most forest managers because they offer simple solutions and maintain control with the management agency. However, over the past 20 years an increasing number of forest managers have come to realize that sophisticated approaches are needed to address more proactively visitor management issues. Some of the more sophisticated approaches include:

1. understanding visitor management issues and opportunities through the use of market research, visitor monitoring and visitor research;
2. influencing visitor expectations and behaviour through marketing;
3. influencing visitor expectations and behaviour through interpretation and education programmes and facilities;
4. modifying the degree and nature of on-site presence of forest management; and
5. encouraging and assisting alternative providers of visitor experiences, such as volunteers and tourism operators (particularly those whose quality is accredited).

Hall and McArthur (1998) critiqued these approaches against the two basic objectives of visitor management in forests – the ability to conserve the forest environment directly and secondly, improve the quality of the visitor experience. The authors used a wealth of case studies throughout their text to demonstrate the basis of their evaluation. The reliance by management to use any one of the techniques was assessed according to whether the technique was used in isolation of other techniques. Table 16.1 provides an account of this critique and shows that strong performers were the use of market research, visitor monitoring and research, marketing, interpretation and education, volunteers and a concentration on accredited tourism operators. Despite the strong performance of these sophisticated approaches, it has been suggested that visitor management is largely reliant on conventional approaches used in isolation of each other (Hammit, 1990; Glasson *et al.*, 1995; Hall and McArthur, 1998). The lack of integration of various approaches means that each one cannot reach its full potential and that many may be working against each other. For example, the regulation of access, equipment and behaviour may be influencing some people to go elsewhere, thereby shifting the problem to a new site. This lack of integration can equally impair the more sophisticated methods. For example: visitor research may not be asking the questions needed to shape a minimal impact education programme, or a marketing programme may not be promoting critical elements that will influence visitor choice of one experience over another.

What is clear from this brief look at conventional and more sophisticated visitor management approaches is that forest managers need to integrate and coordinate the ones they choose to use. An integrated approach calls for a shift in management culture that recognizes the intrinsic balance between people and the places they value and wish to visit. Fundamental to this is the use of management systems or models that attempt to make sense of the complex web of relationships that surround the state of a forest and the visitor experience.

Visitor management models do not, by themselves, provide ready-made solutions. However, as relationships are uncovered and understood, the way is paved for a variety of stakeholders to identify common ground, which, in turn, paves the way for solutions that meet a variety of needs (Hammit and Cole, 1987). Therefore, an integrated approach uses models to help forest managers and their stakeholders establish a clearer understanding of the problems and opportunities, and adapt accordingly.

The Introduction and Failure of Carrying Capacity

One of the most widely recommended models has been Carrying Capacity (Glasson *et al.*, 1995). The application of this model to visitor management came from the desire to compare the amount of visitor activity (number of visitors) with the scale of the impacts generated by tourism (Hall, 1995). The dimensions that Carrying Capacity should address have varied over time. Lime and Manning (1977) suggested the model address biophysical, sociocultural, psychological and managerial dimensions. Glyptis (1991) suggested physical, economic, ecological and social dimensions. Importantly, each dimension is included to produce its own carrying capacity equation, leaving the forest manager with a difficult dilemma – which one to choose. The difficulty in choosing one equation over another has been one of the key stumbling blocks preventing the adoption of Carrying Capacity in Australia (McArthur, 1999).

Comparing the number of visitors with the scale of impact is simple to do but impossible to prove because there are just too many variables. Extensive research into environmental and social impacts has failed to establish predictable links between different levels of use and their impacts (Washburne, 1982; Graefe *et al.*, 1984; Ceballos-Lascurain, 1996). The resilience of a forest to withstand visitor impact changes from day to day and year to year. For example, soil may be able to withstand 10 horses riding over it when it is dry, but may become completely broken up after heavy rain. Further complicating matters is the variation in visitor, particularly their behaviour – one group of visitors may practise minimal impact behaviour and leave no trace, yet the next group may choose to degrade the site as an essential part of their experience. In essence, the simplicity of Carrying Capacity is its greatest strength and its greatest weakness. In the face of expectations for forest managers to use models that take account of complexity and demonstrate flexibility, using Carrying Capacity just is not worth the political flak.

Stankey (1980) suggested three reflections to move beyond Carrying Capacity:

1. the primary concern should be on controlling the impact and its impact, not use *per se*;
2. visitor management should minimize regimentation, favouring less direct methods and then control only when indirect methods fail; and
3. accurate objective data is essential to avoid planning based on coincidences and assumptions.

The difficulties experienced with applying Carrying Capacity and accounting for the subsequent position on levels of use has been the key stumbling block preventing its wider adoption, and today even the architects of the model suggest it be discarded for a more sophisticated and politically sensitive model (Lindberg *et al.*, 1996).

Other Visitor Management Models

As a result of the constant difficulties associated with implementing Carrying Capacity, most visitor managers have shifted their focus from a relationship between levels of use and impact, to identifying desirable conditions for visitor activity to occur in the first place. This shift in thinking has generated more sophisticated models designed to collect information and assist in making subsequent decisions from that information. In essence, these models monitor the state of the visitor experience and the state of the forest that they are experiencing. As soon as the model detects a change in either state a management decision is triggered to alter the way visitors and the site are managed.

Over the past 20 years a number of models have been created and used in this fashion. The Visitor Impact Management Model (VIMM) increased the role of monitoring, identifying the cause of visitor impact and generating strategies to deal with it. The VIMM keeps a relatively conservative focus on minimizing impacts but does at least generate a scenario of management by objectives (Graefe, 1991). Unfortunately, these objectives or desirable conditions are typically narrowly focused on only party of the equation – the state of the environment and, to some extent, the quality of the visitor experience.

The Visitor Experience and Resource Protection Model (VERP) was established to determine the most appropriate visitor experiences based on values and significance, then determines specific conditions for the forest environment to be maintained too (Falvey, 1996). The VERP is added to the VIMM model by applying the designated experiences and forest conditions to a zoning system, then applying a monitoring system to check both are in order. Once completed the VERP is linked to its region's management plan, which is approved by an act of government and thus becomes a legal document. This linkage to a management plan therefore provides consistency and legislative strength.

The Visitor Activity Management Program (VAMP) switched forest managers from a product to market orientation, offering the opportunity for fundamental change in forest management planning systems and the culture of forest management organizations (Graham *et al.*, 1988). It is a planning system that integrates visitor needs with resources to produce specific visitor opportunities. The VAMP was designed to resolve conflicts and tensions between visitors and forest managers, and requires managers to identify, provide for and market to designated visitor groups.

The Limits of Acceptable Change (LAC) model went beyond the VIMM by generating opportunity classes or zones to describe different management approaches to the forest environment, then varying each class to maximize the conservation of the resource and quality of the visitor experience. The LAC model establishes how much change is acceptable, then manages visitors and the forest to keep conditions under these limits (Clarke and Stankey, 1979; Stankey and McCool, 1984; McCool and Stankey, 1992). Specifically, the LAC system first determines what conditions are the most desirable, then monitors the actual situation to determine whether the conditions are within acceptable standards. If they are not, management is then equipped with a logical and defensible case to identify and implement actions to protect or achieve the conditions. One action may be to limit use, but inherent in the model is the generation and evaluation of alternatives, and the monitoring of conditions after their introduction. LAC therefore avoids the use/impact conundrum by focusing on the management of the impacts of use (Stankey *et al.*,1985).

Reflections on Models Produced to Date

The most critical aspect of the development of these models has been establishing stakeholder endorsement and support sufficient to get the models implemented and operational long enough to prove their worth (Prosser, 1986; Hammit, 1990; Glasson *et al.*, 1995; Hall and McArthur, 1998; McArthur, 1999). Stakeholders from the local tourism sector and community are critical to implementing these models. These stakeholders can provide valuable input into desired conditions and acceptable standards, and are usually essential in providing the economic and political support necessary to maintain monitoring programmes and implement management decisions.

The failure to establish sufficient stakeholder support for these models has largely occurred because the culture inherent in the models simply is not attuned to attracting wider stakeholder involvement. Addressing this means overcoming three impediments:

- the use of the terms 'impact' and 'limits' within the title, which stakeholders such as the tourism industry have interpreted as being discouraging to economic growth and prosperity;
- the conventional narrow focus on the condition of the physical environment and to some extent, the nature of the visitor experience; and
- the lack of cooperative involvement of stakeholders (beyond forest managers) in identifying indicators and standards.

Forest managers need to involve stakeholders in the development and implementation of visitor management models. Without this involvement, stakeholders cannot be expected to understand and support the decisions emanating from the model. Without stakeholder support, forest managers can expect the insights and decisions generated by the model to be challenged, particularly if they reveal surprising or controversial implications. The end result may be the overturning of decisions and the forced abandonment of the model itself.

Introducing a New 'Politically Sensitive' Model

In 1996 the Commonwealth Department of Industry, Science and Tourism and the South Australian Tourism Commission initiated a project designed to develop a Limits of Acceptable Change model for Kangaroo Island in South Australia. The large size and diversity of settings and activities of Kangaroo Island meant that a regional application of the LAC was required. Most of the Island is private farms and residential landholdings, but approximately 26% (or 105,000 ha) has been set aside for conservation purposes (PPK Planning, 1993). These areas are largely made up of 13 Conservation Parks, five Wilderness Protection Areas, one National Park and two Aquatic Reserves, which are managed by the National Parks and Wildlife Service, South Australia. Most roadsides act as vegetation corridors between larger areas of remnant vegetation.

Approximately 38% (153,931 ha) of the island has never been cleared, and most of this is found in the western third, which is covered by dry sclerophyll forest and mallee scrub. There are no forestry operations based on native forests and it is illegal to remove any native vegetation without a permit. Nonetheless, a small forestry operation representing some 4000 ha of pine plantation was introduced to Kangaroo Island in 1976 and has been estimated to yield 2.4 million tonnes when harvested in 2001 (PPK Planning, 1993).

The relatively long period of isolation from the mainland has meant that many plants and animals have evolved in isolation from closely related populations on the South Australian mainland. There are some 38 species of endemic plants, and wildlife is prolific and easily viewed. Unlike most of Australia, there are no foxes or rabbits on Kangaroo Island. Manidis Roberts Consultants were commissioned to develop the LAC for Kangaroo Island. The consultants recommended broadening out the concept of the LAC into a new model that addressed the political issues mentioned earlier in the chapter. The system was named a Tourism Optimisation Management Model (TOMM). TOMM is a regional approach to seek and assess solutions to issues that threaten the health of tourism and the resources that tourism depends upon. Specifically, TOMM has been designed to:

- monitor and quantify the key economic, marketing, environmental, socio-cultural and experiential benefits and impacts of tourism activity; and
- assist in the assessment of emerging issues and alternative future management options for the sustainable development and management of tourism activity (Manidis Roberts, 1996).

Most of the components of TOMM are similar to the LAC, but while LAC is strongly focused on the decision-making process, TOMM has a little more emphasis on the contextual analysis and monitoring programme. The differences are largely a reflection of two different political dimensions. The LAC system was designed to serve a single natural area management organization within one land tenure. TOMM was designed to serve a multitude of stakeholders with a multitude of interests, and can operate at a regional level over a multitude of public and private land tenures. The approach used to develop TOMM is a reflection of the two dimensions; the development of the model and the development of an appropriate political culture from which to implement it (Manidis Roberts, 1997). TOMM has three major parts to its structure, contextual analysis, a monitoring programme and a management response.

Part One: Contextual analysis

The contextual analysis firstly involved identifying strategic imperatives, such as current policies and emerging issues. This helped ensure that TOMM could continue to adapt to emerging issues after it had been initiated. The contextual analysis also involved the identification of community values, product characteristics, growth patterns, market trends and opportunities, positioning and branding, and alternative scenarios for tourism in the region. For each scenario potential benefits and costs were forecast and information that might be required to maximize the benefits and minimize the costs identified. This process helped to identify information critical to predicting and managing future tourism activity and was subsequently used to consolidate stakeholder support for the model.

Monitoring programme

The monitoring programme involved the identification of optimum conditions, indicators, acceptable ranges, monitoring techniques; benchmarks, annual performance and predicted performance reporting. Table 16.2 presents the optimum conditions, indicators, acceptable ranges and monitoring techniques that form the basis of Part Two. An optimal condition was defined as a desirable yet realistic status for a sustainable future. Like objectives, they indicate the environment in which tourism should be operating. A draft set of optimal conditions for Kangaroo Island were initially sourced from a set generated through extensive consultation to develop a Sustainable Development Strategy for the island. Manidis Roberts Consultants presented the list to stakeholders at the first of several workshops run to develop the TOMM. Stakeholders at the workshop refined the list and added an additional dimension termed market opportunities. This dimension has strengthened the model's predictive capability, and will assist in improving the match between the tourism market and product, in encouraging greater participation in cooperative marketing and in helping measure the success in attracting the most appropriate market segments. Indicators were also developed with stakeholders at the first workshop. An indicator was defined as a tangible measure of the state of an optimum condition and was therefore used to gain an idea of how close tourism activity was to achieving its optimal conditions. Criteria were listed into a matrix strategically to assess the most worthwhile indicators, and the project's Steering Committee provided feedback on the assessment that resulted in some minor modifications.

After the first workshop the consultants then developed an acceptable range for each of the indicators selected. An acceptable range was defined as an ideal yet realistic range in which an indicator should be performing to reach its desired condition. The initial acceptable ranges for each indicator

Table 16.2. Summary of working components of monitoring systems for Kangaroo Island application of TOMM.

Optimal condition	Indicator	Acceptable range
Economic		
The majority of visitors to KI stay longer than two nights.	Annual average number of nights stayed on KI.	2–7 nights.
The tourism industry is undergoing steady growth in tourism yield.	Annual average growth in total tourism expenditure on KI per number of visitors.	4–10% annual average growth.
The growth of local employment within the tourism industry has been consistent.	Annual average growth in direct tourism employment.	1–3% annual average growth.
Market opportunities		
Operators use market data to assist in matching product with market segment opportunities.	Number of operators using market data in TKI and operator plans.	50–100% of operators.
There is integration of business and regional, state and national tourism marketing programmes for KI.	Number of cooperative marketing campaigns such as brochures and advertisements.	50–100% of operators.
A growing portion of visitors come from the cultural/environmental segments of the domestic and international markets.	Proportion of visitors that match ATC cultural/ environmental segmentation profile.	60–80% of total visitors to KI.
	The number of visits to Kangaroo Island.	0–7% annual growth in the number of visits.
Environmental		
The majority of the number of visits to the island's natural areas occurs in visitor service zones.	The proportion of KI visitors to the island's natural areas who visit areas zoned specially for managing visitors.	85–100% of visitors.
Ecological processes are maintained or improved (where visitor impact has occurred) in areas where tourism activity occurs.	Net overall cover of native vegetation at specific sites.	0–5% increase in native vegetation base case.
Major wildlife populations attracting visitors are maintained and/or improved in areas where tourism activity occurs.	Number of seals at designated tourist site. Number of hooded plover at designated tourist site. Number of osprey at designated tourist site.	0–5% annual increase in no. sighted.
The majority of tourism accommodation operations have implemented some form of energy and water conservation practice.	Energy consumption/visitor night/visitor. Water consumption/visitor night/visitor.	3–7 kW. 20–40 l of water.

Experiential		
Tourism promotion of visitor experiences at Kangaroo Island's natural areas is realistic and truthful to that actually experienced by most visitors.	Proportion of visitors who believe their experience was similar to that suggested in advertisements and brochures.	85–100% of visitors.
The KI visitor experience is distinctly different from other coastal destinations in Australia.	Proportion of visitors who believe they had an intimate experience with wildlife in a natural area.	70–100% of respondents.
The majority of KI visitors leave the Island highly satisfied with their experience.	Proportion of visitors who were very satisfied with their overall visit.	95–100% of respondents.
	Proportion of visitors who were very satisfied with interpretation provided on a guided tour.	90–100% of respondents.
Socio/cultural		
Residents feel they can influence tourism-related decisions.	Proportion of residents who feel that the local community is capable of influencing the type of tourism on KI.	70–100% of residents.
Residents feel comfortable that tourism contributes to a peaceful, secure and attractive lifestyle.	Number of petty crime reports committed by non-residents per annum.	10–25 crime reports per annum residents.
	Number of traffic accidents involving non-residents per annum.	50–80 vehicle accident reports per annum.
	Proportion of the community who perceive positive benefits from their interactions with tourists.	70–100% of respondents.
Residents are able to access nature-based recreational opportunities that are not frequented by tourists.	Proportion of residents who feel they can visit a natural area of their choice with very few tourists present.	80–100% of respondents.

Source: Manidis Roberts (1997: 26).

were developed using previous monitoring and research where available, then observations and estimations from those with experience and expertise in the given field where data was unavailable. It was accepted that some of the acceptable ranges would need to be adjusted if monitoring data from the TOMM suggested the initial range had been unrealistic – this was seen as the practical reality of implementing a new model.

A monitoring programme was then developed to collect information about how close the condition of each indicator was to its acceptable range. The monitoring programme was defined as the way in which information is collected and stored over a period of time. Most of the monitoring techniques were suggested by stakeholders in a second workshop with the consultants. The potential monitoring techniques were then assessed by the consultants using similar performance measures as the indicators, such as reliability, cost and the ability to demonstrate causal-effect relationships. The preferred methods were those coming from existing systems and initiatives, followed by those that could be developed to satisfy other information needs, followed by those that could only be applied to this project. The two new and most substantial monitoring methods developed for this application of a TOMM were a continual visitor exit survey and a tourism operator survey. Most of the environmental monitoring techniques were additions or adaptations to existing programmes. A benchmark was developed for each indicator selected. A benchmark was defined as an indicator's point of reference against which to compare new monitoring data. The initial benchmarks for each indicator were developed using the same sources as the acceptable ranges. The benchmarks represented the first tangible expression of what the TOMM reporting would look like. The data from the model's first year's operation may well be more accurate and suggest some modification of some benchmarks.

The principle data generated from the model come from the monitoring programme. These data reflect the annual status of each indicator. The inclusion of acceptable ranges helps provide a management context to the data by presenting the annual status in terms of how close it is to the acceptable range. The annual status thus becomes annual performance. The closer the state of the annual data is to the acceptable range, the closer tourism activity is to reaching optimal performance. Annual performance is presented in two ways. Reporting Tables provide a 'quick and dirty look' and Reporting Charts provide the opportunity for trends and interpretation of data. The Reporting Table is a simple table instantly conveying to any reader whether an indicator is within its acceptable range or not. Each indicator is given a tick or cross depicting whether the annual data is within or outside the acceptable range, or whether it has reached optimal performance, which is beyond the most favourable end of an acceptable range. The Reporting Table is supplemented by a report chart for each indicator. Each reporting chart presents the indicator's benchmark, acceptable range, annual past performance, predicted performance and qualitative comments about the data that may provide useful context.

Part Three: Management response

The third part of TOMM involves the identification of poor performing indicators, the exploration of cause/effect relationships, the identification of results requiring a response and the development of management response options. The response should result in the development of management options to address poor performing indicators. This in turn should assist in achieving the desired outcomes for tourism activity. The first stage in the response mechanism is to identify annually which indicators are not performing within their acceptable range. This involves reviewing the report charts to identify and list each indicator whose annual performance data is outside its acceptable range. It also involves identifying the degree of the discrepancy and whether the discrepancy is part of a longer-term trend. The trend is determined by reviewing previous annual data that has been entered onto the report charts. A qualitative statement is then entered under the degree of discrepancy. An example for Kangaroo Island may be visitor's average energy consumption per night. The upper range for this indicator is 0.5 kW. Consumption over the three previous years may have been 0.3, 0.35 and 0.5 kW. The current year being assessed may be 0.7 kW. The indicator would thus be listed as not having performed within its acceptable range. The discrepancy would be 0.2 kW. The trend would be 'a gradual increase that does not appear to be tapering off'.

The second stage in the response mechanism is to explore cause/effect relationships. The essential question relating to cause and effect is whether the discrepancy was principally induced by tourism activity. Many indicators are subject to other effects such as the actions of local residents, initiatives by other industries, and regional, national or even global influences. Many cause/effect relationships are relatively easy to identify, particularly in remote and sparsely populated areas. Effects that might display a direct relationship to tourism activity on Kangaroo Island might be the direct comments about the quality of interpretation experienced. Relationships that may not be relevant to tourism activity on Kangaroo Island might be a national pilot strike which prevented a proportion of visitors flying to the Island, or a change in sea currents that moved seal food (and thus the seals) away from the waters immediately surrounding Kangaroo Island.

The third stage would simply involve nominating whether a response is required. Specific choices for the response could include a tourism-oriented response, a response from another sector, or identification that the situation is beyond anyone's control. The final stage is to develop response options. These response options could take one of three forms. The first form would be for indicator results requiring a response from a non-tourism sector. This would involve identifying the appropriate body responsible, providing them with the results and suggesting a response on the matter. The second form would be for indicator results that were out of anyone's control. In this instance, no response would be required. The third and most likely form would be for indicator results requiring a response from the tourism sector. This would involve generating a series of management options for consideration. These options could include additional research to understand the issue, modification to existing practices, site-based development, marketing and lobbying.

After the tourism-related options are developed the preferred form should be tested using the model. This would involve brainstorming how the option might influence the various indicators. This requires the re-use of the predicted performance and management response sections of the model. Once several years of data are collected it may be worth transferring the model to a simple computer program. This would streamline the reporting, predicting and testing of options. The final application of the model is to test potential options or management responses to a range of alternative scenarios. The first form of testing for application to Kangaroo Island was the performance of a sample of individual indicators. The second form of testing the model's performance was against several potential future scenarios for Kangaroo Island already developed and presented in the contextual analysis. This helped to ensure that the model would have some degree of predictive capability.

Implementation of the Kangaroo Island TOMM

The final version of the model was launched in early 1997 (McArthur and Sebastian, in press). The first task achieved in implementing the model was the formation of an Implementation Committee in mid-1997. In late 1997 the Committee organized the design and production of the principal monitoring component of the model – a visitor survey to collect details of the type of visitors and their impressions of various aspects of the Island and its tourism product. The survey was implemented in late 1997 and the first data produced in early 1998 (McArthur, 1999). During 1998 some of the environmental monitoring was established (particularly the condition of seals at Seal Bay), along with monitoring of visitors to various natural area destinations (McArthur, 1999). Collation and presentation of this data in the TOMM reporting format has been held off until more of the monitoring is established in late 1999.

In mid-1998 the Management Committee gained access to further external funds to employ a project manager to coordinate the model. Substantial cash and in-kind support from each of the partners of the Implementation Committee, in conjunction with Commonwealth funding, has created a resource pool of Aus$260,000 for collaborative implementation of the TOMM during 1999–2001 (Twyford *et al.*, in press). The TOMM for Kangaroo Island is considered relatively expensive to run, though there are very few published examples of definitive budgets and operating expenses for visitor management models to benchmark against (McArthur, 1999). It will cost the three major participating organizations Aus$70,000 per annum to run the TOMM over its first 5-year period (Manidis Roberts, 1997). An analysis of funding sources was undertaken by the consultants, who identified short-term funding from the three organizations and a one-off grant would be sufficient to run the TOMM for its first 2–3 years. A minimum of 3 year's operation is required to run the model long enough to evaluate its performance and determine its long-term resourcing and support. The consultants identified that the best medium-term funding would come from the combination of a local government rates increase and a departure tax for visitors. To be palatable, the consultants devised a break-up of funds between the running of the TOMM and the development of infrastructure such as roads. In year four it was also proposed to build in a user-pays system whereby new tourism developments over $20,000 would be required to use and pay for the use of the TOMM to predetermine their value against the optimum conditions.

The Implementation Committee is now looking to accelerate progress, as it is widely recognized that the more implementation achieved in the short term, the more support would be generated to assist with full implementation and ongoing support. It has been proposed that some of the monitoring techniques that merely reflected reworking existing data should be concentrated on, to act as an incentive to tackle the more involved techniques (McArthur, 1999). The Committee is also seeking to intensify the marketing of TOMM (McArthur, 1999). The most challenging targets of this marketing are the councillors and key transport and tourism operators. The support of the councillors is regarded as critical to introducing a user-pays financial resourcing system, in the form of a visitor and rates-based levy. While direct marketing (face to face) was recommended in the short term, marketing to the broader community was felt to be the most effective means of achieving sustainable councillor support into the medium term. Getting schools involved in TOMM through the provision of user-friendly information about it and the invitation to become involved in the monitoring is also seen as a valuable means of gaining community support, and thus councillor support.

It is planned to market TOMM further through the development of a Web-site that would provide up-to-date information on results to date as well as background information about the development and implementation of the model. The Web-site is planned to be produced in early 2000. At this stage there is considerable optimism that TOMM can be implemented. This appears to be fuelled by the belief among the Implementation Committee that TOMM represents a real opportunity to break out of an apparent tradition of apathy, help the Island's tourism industry grow and arrest cumulative deterioration in what makes Kangaroo Island special. Whether this is realized is questionable given the poor track record in implementing other initiatives. The general sentiment among most stakeholders involved in the model's development can be loosely quoted as 'if we don't make TOMM work, we'll lose the Island magic'.

Implementation elsewhere

The TOMM has only been in existence for 1½ years so examples of its application are few and far between. To date there are only two known full applications (both in Australia) and several partial applications (in Canada). However, there are at least four or five applications being considered for high-profile tourism sites in Australia and Canada by protected-area managers and tourism authorities. Table 16.3 presents known and planned applications of TOMM in 1999.

Limitations to the TOMM

The most obvious limitation of the TOMM is its sheer size compared with any other model reviewed. TOMM works at the regional level and therefore covers the widest range of environments, tenures, land uses, issues and stake-

Table 16.3. Actual and proposed applications of the Tourism Optimisation Management Model across the World.

Australian and Canadian applications	Proposed applications in Australia and Canada
In Australia: Being implemented at Kangaroo Island, South Australia, Australia, by the South Australian Tourism Commission, Department of Environment and Natural Resources and Tourism Kangaroo Island (McArthur and Sebastian, in press). Implemented at Dryandra Woodland Reserve in Western Australia by the Department of Conservation and Land Management (Moncrief, in press), with first year's data now being compiled.	Proposed for a range of key protected area sites across the State of New South Wales (New South Wales National Parks and Wildlife Service, 1997). Being considered by Parks Victoria for Port Campbell in Victoria (McArthur, 1999). Being considered by Tourism NSW and Manly Council for Manly or far northern New South Wales (McArthur, 1999). Being considered for a series of regional applications in South Australia such as Coorong National Park (McArthur, 1999). Being considered for the entire of South Australia (McArthur, 1999).
In Canada: Partially implemented by Parks Canada at Lake Louise (Banff National Park) to give context to a transportation study (McArthur, 1999). Partially implemented by Parks Canada at Aulavik to accommodate dimensions such as commerce, ecology, geography, economics, sociology and anthropology. The project is currently developing indicators and linking in existing monitoring (McVetty and Wight, 1998).	Being considered by Parks Canada for Gwaii Haanas, a fragile rainforest park reserve with aboriginal co-management, and with competing tourism, logging, and preservation pressures (McArthur, 1999). Being considered by Parks Canada for Bathurst Island (a remote Arctic area with co-management soon to be negotiated as a park (McArthur, 1999)).

holders. This scope means that TOMM offers a 'taste of everything' rather than a moderate understanding of some things. It therefore needs to be recognized as an indicative tool requiring back-up research where relationships are particularly complex. In addition, the three dimensions added to the TOMM (market, economic and socio-cultural) effectively add a lot more work to managing data and the stakeholders who collect it. The increased range of stakeholders also means much greater coordination and much greater political skill to deal with such a diversity of interests and expertise.

Another limitation is the same one that occurs for the VIMM and LAC models – selecting the right indicator to suggest the state of the optimal condition. An obvious trade-off must be made between having sufficient indicators to reliably know the state of the optimal condition versus the cost of managing the data from each indicator. The wider range of fields in a TOMM generates a wider range of optimal conditions that, in turn, generates a wider range of indicators. This puts pressure on stakeholders to limit the range of indicators per condition to keep the model manageable, but in turn limits the ability of each condition to be adequately represented. The TOMM also shares the potential for subjectivity when choosing the acceptable range and benchmark. While a range provides greater flexibility than a limit, the benchmark remains a fixed value whose initial nomination is typically generated by a calculated guess. The initial predicted trends must also be guesstimates until 2 or 3 years of data have been collected.

Conclusions

Long-held concerns by forest managers about visitor impact in natural areas has in recent years broadened into the community and parts of the tourism industry. These groups have equally significant stakes in visitor impact relationships, and must now be integrated into research, monitoring and management responses. The Limits of Acceptable Change system has to date provided one of the most comprehensive approaches to understanding these relationships and responding with appropriate management actions. However, there is a need for a model that reflects a broader range of values, has a broader scope and thus appeases a broader range of stakeholders.

The TOMM has been developed to help generate tangible evidence that the viability of the tourism industry is dependent upon the quality of the visitor experiences it generates, and the condition of the natural, cultural and social resources it relies on. However, the author does not argue that the TOMM is the ideal model to replace past models, nor is it argued that the TOMM need be delivered with such a strong tourism emphasis. The philosophy of the TOMM is particularly valuable for use in forests where values are diverse and thus competition for different outcomes is intense. The three examples where the TOMM has already been implemented (Kangaroo Island, Dryandra Woodland and Banff National Park) suggest that there is real merit in not only using the TOMM to manage visitors in forests, but in integrating the essence of the TOMM into broader environmental management of forests. There is no reason why the basic development process of a TOMM could not be applied to a general model for monitoring and managing forests. Indeed, forest managers across Australia have already jointly identified indicators and targets for the conservation and sustainable management of temperate and boreal forests, as part of the 'Montreal Process' (Department of Primary Industries and Energy, 1997). This initiative will provide a tool for assessing national trends in forest conditions and management, and provide a common framework for describing, monitoring and evaluating progress towards sustainability. While this initiative and the TOMM are similar in their first two stages (context analysis and monitoring), the TOMM sets up a simple performance standard (acceptable range) and a simple reporting system that makes it accessible to a wider range of stakeholders. In addition, the TOMM has built on a follow-on decision-making system that further involves the same stakeholders.

The implementation of a model such as TOMM to manage visitors to forests, or the implementation of a model similar to the TOMM into broader forest management, could greatly assist to shift the culture of various stakeholders to one where people understand each other's values and needs, and the intricate dependency relationships between each value and need. Anything which can help achieve such an outcome is surely worth considering.

References

Ceballos-Lascurain, H. (1996) *Tourism, Ecotourism and Protected Areas, The State of Nature-based Tourism Around the World and Guidelines for its Development.* The World Conservation Union, pp. 130–136.

Clark, R.N. and Stankey, G.H. (1979) The recreation opportunity spectrum: a framework for planning, management and research. USDA Forest Service General Technical Report PNW-98. Portland, Oregon.

Department of Primary Industries and Energy (1997) Australia's first approximation report for the Montreal Process, Commonwealth of Australia, Canberra.

Falvey, D.A. (1996) Seeking public input for the future of Zion National Park. *Zion Newsletter*, no. 1, Sept. edition. US National Parks Service, Zion National Park.

Glasson, J., Godfrey, K., Goodey, B., Absalaom, H. and Van Dert Borg, V. (1995) *Towards Visitor Impact Management – Visitor Impacts, Carrying Capacity and Management Responses in Europe's Historic Towns and Cities.* Averbury, Aldershot, pp. 27–156.

Glyptis, S. (1991) *Countryside Recreation.* Longman Group, Harlow, pp. 148–152.

Graefe, A.R. (1991) Visitor impact management: an integrated approach to assessing the impacts of tourism in national parks and protected areas. In: Veal, A.J., Jonson, P. and Cushman, G. (eds) *Leisure and Tourism: Social and Environmental Change, Papers from the World Leisure and Recreation Association Congress, Sydney, Australia.* University of Technology, Sydney, pp. 74–83.

Graefe, A.R., Vaske, J.J. and Kuss, F.R. (1984) Social carrying capacity: an integration and synthesis of twenty years of research. *Leisure Sciences* 6, 395–431.

Graham, R., Nilson, P. and Payne, R.J. (1988) Visitor management in Canadian national parks. *Tourism Management*, March, 45–62.

Hall, C.M. (1995) *An Introduction to Tourism in Australia: Impacts, Planning and Development.* 2nd edition. Longman Australia, Melbourne.

Hall, C.M. and McArthur, S. (1996) *Heritage Management in Australia and New Zealand – The Human Dimension.* Oxford University Press, Melbourne, pp. 37–51.

Hall, C.M. and McArthur, S. (1998) *Integrated Heritage Management – Principles and Practice.* The Stationery Office, Norwich, pp. 107–138.

Hammitt, W.E. (1990) Wild land recreation and resource impacts: a pleasure-policy dilemma. In: Hutcheson, J.D., Noe, F.P. and Snow, R.E. (eds) *Outdoor Recreation Policy – Pleasure and Preservation.* Greenwood Press, New York.

Hammitt, W.E. and Cole, D.N. (1987) *Wildland Recreation, Ecology and Management.* John Wiley, New York, pp. 195–212, 327–337.

Kuss, F., Graefe, A. and Vaske, J. (1990) *Visitor Impact Management, A Review of Research*, Vol. 1. National Parks and Conservation Association, Washington DC, pp. 1–7, 231–243.

Lime, D.W. and Manning, R.E. (1977) Crowding and carrying capacity in the national park system – towards a social science research agenda. In: *Crowding and Congestion in the National Park System – Guidelines for Management and Research.* Department of Forest Resources and the Minnesota Agricultural Research Experiment Station, University of Minnesota, pp. 50–65.

Lindberg, K., McCool, S. and Stankey, G. (1996) Rethinking carrying capacity. *Annals of Tourism Research* 24, 461–465.

Manidis Roberts Consultants (1996) Tourism optimisation management model for Kangaroo Island, a model to monitor and manage tourism. Consultation Draft, South Australian Tourism Commission, Adelaide, pp. 1–35.

Manidis Roberts Consultants (1997) Developing a tourism optimisation management model (TOMM); a model to monitor and manage tourism on Kangaroo Island. Final Report, South Australian Tourism Commission, Adelaide, pp. 1–75.

McArthur, S. (1999) Visitor management in action – an analysis of the development and implementation of visitor management models at Jenolan Caves and Kangaroo Island. PhD thesis, University of Canberra, Canberra, Australia.

McArthur, S. and Sebastian, I. (in press) Visitor impact management models – who's done or doing what in Australia? In: *Proceedings of 1998 Annual Ecotourism Conference*, held at Margaret River, Western Australia. Ecotourism Association of Australia, Brisbane.

McCool, S. and Stankey, G. (1992) Managing for the sustainable use of protected wildlands: The limits of acceptable change (LAC) framework. Paper presented to the Fourth World Congress on National Parks and Protected Areas, Caracas, Venezuela.

McVetty, D. and Wight, P. (1998) Integrated planning to optimise the outcomes of tourism development, the case of Aulavik National Park and Banks Island, NWT. In: *Proceedings of the Travel and Tourism Research Association 1998 Annual Conference.* The Travel and Tourism Research Association (Canada Chapter), Toronto, Ontario.

Moncrief, D. (in press) Managing the impacts of Tourism and Recreation at Dryandra Woodland. In: *Proceedings of 1998 Annual Ecotourism Conference*, held at Margaret River, Western Australia. Ecotourism Association of Australia, Brisbane.

NSW National Parks and Wildlife Service (1997) Draft nature tourism and recreation strategy. NSW National Parks and Wildlife Service, Hurstville, pp. 36–48.

PPK Planning (1993) *Kangaroo Island Planning Strategy: Statement of Investigations.* Unpublished report for District Council of Dudley and the District Council of Kingscote Plan Amendment Reports.

Prosser, G. (1986) The Limits of Acceptable Change: an introduction to a framework for natural area planning. In: *Australian Parks and Recreation.* Royal Australian Institute of Parks and Recreation, Canberra, Autumn edition, pp. 5–10.

Stankey, G.H. (1980) The application of the carrying capacity concept to wilderness and other low-density recreation areas. In: Robertson, R.W., Helman, P. and Davey, A. (eds) *Wilderness Management in Australia: Proceedings of Symposium.* Canberra College of Advanced Education, Canberra, pp. 150–177.

Stankey, G.H. and McCool, S.F. (1984) Limits of Acceptable Change: a new Framework for managing the Bob Marshall Wilderness Complex. *Western Wildlands,* Fall edition.

Stankey, G.H., Cole, D.N., Lucas, R.C., Peterson, M.E. and Frissell, S.S. (1985) The Limits of Acceptable Change (LAC) system for wilderness planning. USDA Forest Service General Technical Report INT-176. US Dept Agriculture & US Forest Service, Ogden, Utah.

Twyford, K.L., Vickery, F.J. and Moncrief, D. (in press) Development and application of Tourism Optimisation Management Models (TOMM) at Kangaroo Island and the Dryandra Woodland. In: Worboys, G., De Lacy, T. and Lockwood, M. (eds) *Protected Area Management in Australia.* Environment Australia and NSW NPWS, Hurstville.

Washburne, R.F. (1982) Wilderness Recreational Carrying Capacity: are numbers necessary? *Journal of Forestry* 80, 726–728.

17

Implementing Environmental Management Systems in Forest Tourism: the Case of Center Parcs

Barry Collins

Introduction

In his book *Holland and the Ecological Landscapes*, Alan Ruff identified 1968 as the beginning of a new environmental movement. The new landscapes, based on a philosophy of man's relationship to nature, established new images and new techniques, and set standards for landscape design and management which are now the accepted norm in a far more environmentally aware continent.

It is no coincidence that the Center Parcs concept was developed at the same time, in the same country. It responded to and helped to shape the same philosophy, that man needs contact with nature for his well-being and that development and land management needed a completely new outlook, based on ecological principles.

Center Parcs was unique in responding commercially by offering people short breaks close to nature in a forest setting in a manner which embodied the very essence of these new principles. Often seen as an idea before its time, Center Parcs was, in fact, a generator of ideas, setting new standards of environmental awareness and care both in its philosophy and in its day-to-day operations, thus pioneering the reality of sustainable tourism.

Today Center Parcs is one of the leaders in the UK short break holiday market. The principal element of the Center Parcs concept is to allow their guests to escape from the pressures of modern day living by enjoying close contact with nature with all its restful and restorative qualities. In order to realize this concept there are no boundaries between 'Nature areas' and 'People areas'. To achieve this, and stemming from the original design principles, management for wildlife extends right up to the guest villas with each villa patio being a vantage point for a wealth of wildlife, from birds to wildflowers, dragonflies to wild deer.

The size of the village is important with each being approximately 400 acres allowing room for the development of a mix of natural habitats in the vicinity of each villa. Less than 10% of the area of each site is developed. The mix of wildlife habitats adds to the landscape aesthetics resulting in a landscape attractive to both people and wildlife. The principal habitat components are a network of streams and waterways, integrated through the coniferous woodland, with areas of acid grassland and herb-rich glades and rides, and under planting with a range of deciduous and mixed woodland species.

From its arrival in the UK in 1987, Center Parcs recognized that in order to achieve the optimum mix of landscape aesthetics and biodiversity, a detailed Environmental Management System (EMS) was required. In order for the EMS to function to its optimum, a forum of significant expertise was established and a process of consultation introduced. This forum itself developed over the following 12 years and today is represented by Center Parcs' own team of specialists comprised of Chartered Landscape Architect, Ecologist, Environmental Manager and Countryside Managers. This team is supported by expert input from one of the UK's leading ecologists, Dr C.W. Gibson, who has been working with Center Parcs for more than 10 years. Finally this forum is supplemented by advice and recommendations from specialist environmental organisations, for example The County Wildlife Trusts interact at a local level, with English Nature and the Countryside Agency providing advice on a national level.

The resulting EMS is a detailed Forest Management Plan containing prescriptions for all desired landscape characteristics. This interacts with an extensive ecological monitoring programme.

The environmental management system is therefore not only comprehensive but also highly applicable to the conservation of habitats and species on both a local and national level. The history of the EMS is one of constant evolution where the status and knowledge of species and habitat requirements has expanded on an annual basis focused not only on the village landscape, but also the conservation strategies of the UK as a whole. Two key milestones in the development of Center Parcs' Environmental Management System were achieved over 1998 and 1999. Firstly, the results of the detailed annual Ecological Monitoring Studies coupled with the progress of the Forest Management Plan on each village provided the opportunity to develop an Action Plan for Biodiversity containing all the elements of a complete management system. This plan which supplemented and succeeded the original landscape EMS, the Forest Management Plan, was launched on each village in 1998.

Secondly, as a responsible organization Center Parcs recognized the need for an environmental management system that encompassed all the environmental aspects of its operation. The development of this holistic EMS reached its pinnacle in June 1999 with the award of ISO14001, the

International Standard for Environmental Management Systems, Center Parcs being the first leisure organization in Europe to achieve this accreditation. This case study focuses on one part of this accredited EMS, namely Center Parcs' commitment to Biodiversity Management, which represents approximately 20% of the ISO14001 system.

EMS for Wildlife Conservation at Center Parcs

Wildlife Conservation represents one of the most important aspects of modern life with the protection and enhancement of the natural environment one of the key elements of sustainable development. It is therefore essential that landowners and conservation bodies, as guardians of our country's natural history, care for the environment and strive to reduce the impact that modern living has upon it. Center Parcs has accepted not only this responsibility but recognizes that in order to achieve this duty, a complete management ethos is required.

The International Standard ISO14001 provides a template for achievement. The major cause of the failure of any environmental management system is that one essential element of the management cycle (see Fig. 17.1) is either inadequately specified or simply incomplete. The advantage of joining a compliance scheme is the specific demands of the compliance scheme itself and the detailed interrogation of the system through an external audit process. Ultimately all environmental management systems (EMS) require a complete quality management process. This process is often based on the management philosophy known as the Deming Cycle, which is purely and simply sound management practice (Fig. 17.1).

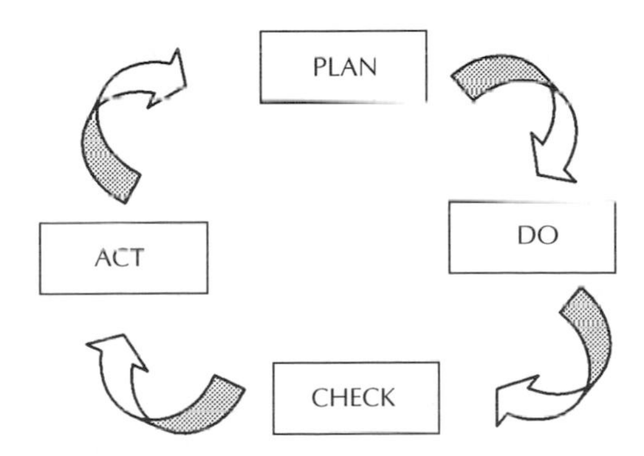

Fig. 17.1. A management system cycle.

Success using this philosophy can be demonstrated by breaking down the Center Parcs Forest and Biodiversity Management Plan into these core elements:

Plan

At this stage an organization should establish a policy appropriate to the nature, scale and environmental impacts and opportunities of its activities. The policy should provide the framework for setting and reviewing the objectives and targets of the EMS. In an organization as large as Center Parcs this focuses on several landscape aspects such as site selection, development activities and landscape establishment. However, the focus of this case study is on site management and the objectives can be summarized as 'to preserve and enhance the biodiversity of the village landscape in order to reinforce the Center Parcs guests perception of truly being at one with nature'.

In order to achieve this objective a clear understanding of the natural features typical of the local areas and their status is required and from this specific targets set. These targets must be measurable and each Center Parcs village has 30 or more targets established, one example being 'to maintain scrub and edge management rotation to ensure suitable breeding habitats for a target of at least 15 pairs of linnet and 20–25 pairs of songthrush'. The key to establishing these targets is the detailed annual ecological monitoring programme, which has been in operation for the past 9 years. Center Parcs villages are amongst the most studied wildlife sites in the UK as a whole and this monitoring system is central to the ecological achievements of the organization.

Do

Center Parcs implements the actions required to achieve its biodiversity targets through the Annual Forest Work Plan. This detailed plan contains all the essential elements of an implementation plan detailing where action is required, the quantity, the optimal timing and details the standard required and benefits upon completion. Two principal documents are the source of the annual work plan. Firstly, the Forest Management Plan identifies the desired landscape characteristics and prioritizes areas furthest from these end goals. This plan also contains the biodiversity mission statement, objectives and targets. Second, the previous year's Ecological Monitoring Report highlights actions required to achieve the biodiversity targets and provides a quantified assessment on the status of the village wildlife and habitats.

This short chapter summarizes the 'DO' part of the cycle. However, the implementation and operation of the management system requires detailed input. As mentioned above, the plan should prioritize actions but should go further in identifying resources required as well as the specialized skills and equipment necessary. To achieve this Center Parcs ensures that all members of both its management and landscape teams are trained in the Forest and Biodiversity Action Plans. The organization goes further by ensuring all 3500 of its UK employees spend a part of their company induction learning the importance that wildlife conservation represents to the Center Parcs concept.

Check

Whilst all elements of the management cycle are critical in achieving success it is the element of auditing that is most often neglected, often resulting in failure to meet objectives. Center Parcs is particularly proactive in this area, monitoring and measuring progress to target across the life span of each annual forest work plan. Two audit systems are employed. The first is a review of the actions taken against the annual forest work plan, which takes the form of a quarterly review where all the key players managing the system appraise the work completed to date. This also allows the plan to be adjusted should the need for instant action be identified during the Ecological Monitoring Programme. For example in 1998 the rare falcon, the hobby (*Falco subbuteo*) was observed 'prospecting' for a nest site in a woodland compartment due for thinning. Works were immediately suspended until it was identified as safe to return.

The second audit function is the comprehensive Ecological Monitoring Programme conducted on an annual basis. This programme is designed to allow the assessment of the health and status of the village habitats and species, the biodiversity, and identifies any risks developing to 'desired' habitats, reliant wildlife and the landscape character of the village. The programme assesses floral development on a 3-yearly cycle, concentrating on a third of the village per annum, and primarily focuses on the status of desired habitats, habitats typical of the local area as identified by natural area profiles and therefore those most suitable for supporting a comprehensive diversity and density of wildlife. The programme further focuses on aquatic habitats, assessing both floral and faunal interest on an annual basis. Breeding birds surveys are also conducted annually along with established national survey schemes for bats, birds and butterflies. Finally the status of the landscape is verified by specific annual studies of invertebrate groups designed to investigate both positive and negative trends identified by the above research.

In conclusion, all elements of the annual forest work plan and the biodiversity action targets are audited at frequent intervals so maintaining the momentum which has led to the environmental success with which Center Parcs is synonymous.

Act

This stage of the management cycle is as vital as the others and is often referred to as 'closing the loop'. At Center Parcs there are several actions in place to ensure the continued suitability, adequacy and effectiveness of the EMS.

With regard to the Forest and Biodiversity Management Plan, the principal process that reviews and recommends action is the annual Ecological Monitoring Report. This detailed document identifies the status of the village biodiversity, assesses the affects of management and most importantly recommends action for the future. The success of Center Parcs' EMS lies in taking these results and transforming them into management recommendations, which then form a principal element of the proceeding annual forest work plan. The strength of these recommendations lies in the expertise and effort employed in their development. These recommendations come in two forms. First, they identify the steps required in order to maintain and enhance species and habitat diversity. Second, in order to preserve the biological status quo, Center Parcs has developed its own species protection protocol known as Biodiversity Sensitive Areas (BSA).

The management recommendations are derived from both internal and external specialist audit functions mentioned previously. Also, and most obvious, the ecological monitoring results themselves. The end result is that the Center Parcs villages continue to develop as important nature reserves. This also ensures that available management resources are targeted at the practices that will maintain the highest ecological return.

The Biodiversity Sensitive Area (BSA) protocol is in place to ensure protection of the nationally important or legally protected species of each Center Parcs village. The system operates by the declaration of areas of the village, important for rare species as 'no-go areas'. Basically no landscape work that could affect sensitive flora and fauna can take place within the protection zone without consultation with Center Parcs' own specialists. The BSAs are clearly identified to Center Parcs' landscape teams by large-scale colour plans and one-to-one training conducted annually.

Such a strict system was called for because of the wealth of rare and important species that have been attracted to Center Parcs villages on an annual basis. These important species can be difficult to manage, because they can be legally protected or difficult to identify due to their size or likeness to more common species. The very scarcity of these species represents a challenge to any countryside manager, as it is easy to damage or disturb some of these sensitive species. This can occur even when routine good management practices such as woodland management, mowing, chemical applications, tree planting, fertilizer application, grass seed mixtures, to name a few, are carried out simply due to a lack of understanding of species requirements or even their precise location.

As responsible countryside managers, Center Parcs has made a corporate commitment to the preservation of these plants and animals across all of its villages by the Biodiversity Sensitive Area protocol. This detailed review not only allows Center Parcs the opportunity to assess the effectiveness of its actions, but also to audit progress to its biodiversity action targets annually. The information and conclusions generated are then incorporated into the company's annual forest and biodiversity plans and the management cycle commences once more.

The Results: Center Parcs' Landscape and Ecological Achievements

Whilst a comprehensive environmental management system involves considerable time and effort to ensure its maintenance, Center Parcs is an ideal model to demonstrate the benefits of adopting this approach. Ultimately Center Parcs is in the enviable position of being unique in possessing a biodiversity action plan which has passed the stringent requirements of ISO 14001 certification. This accreditation demonstrates the quality of the Center Parcs EMS; however, the results on the ground really justify the enormous effort the organization has put into its care for the environment.

It is difficult to appreciate, when reading these highlights, the poor biological status of each of these coniferous woodlands prior to the arrival and development of the Center Parcs villages. Indeed one of the principal objectives of the development element of the Center Parcs EMS is that it should avoid areas of great wildlife or landscape value. All of their villages have been developed on areas of commercial coniferous woodlands, which are typically areas low in biodiversity. The following list mentions some of the biodiversity successes of Center Parcs UK villages:

Sherwood Forest

- Between 50 and 100% of the Nottinghamshire breeding population of crossbill.
- A total of 118 species of bird now recorded.
- Six Biodiversity Action Plan (B.A.P.) birds breed – bullfinch, linnet, reed bunting, spotted flycatcher, turtle dove and songthrush.
- The B.A.P. short-list bat, pipistrelle, thrives alongside five other species. Annually a nursery colony of up to 400 pipistrelle are present. This bat is suffering a 50–70% decline across Europe.
- 383 species of flora identified.
- 248 species of moth, of these 19 are locally or nationally scarce.
- 102 species of wild bees and wasp, five nationally scarce including a Red-Data Book nomad bee.
- 485 species of *Diptera* (flies), 17 of local or national scarcity.
- Including the nationally vulnerable empid fly, *Platypalpus infectus*.

Longleat Forest

- Over 340 species of flora recorded to date including the locally rare birds-foot, *Ornithopus perpusillus.*
- 48 species of bird breed on the village, 100% increase over pre-construction levels. Including seven breeding pairs of the rare firecrest in 1999 alone.
- The man-made lakes and ponds declared as locally important for conservation in 1997. (Three years after construction.)
- 266 species of moth recorded including 13 locally or nationally scarce species.
- 97 species of wild bees and wasps with nine locally or nationally scarce.
- 387 species of *Diptera* (flies) including ten species of local or national importance. The Red-Data Book syrphid fly, *Pelecocera tricincta,* being one.
- 15 species of dragonfly recorded to date including locally scarce migrant hawker.

Elveden Forest

- Over 400 species of flora recorded to date including 26 nationally rare species, the highlights being:

  1. The nationally endangered fingered speedwell, only one of four sites in the country.
  2. The nationally rare wall bedstraw, only one of three sites in Suffolk.
  3. The largest stand of white horehound, nationally rare, in the county of Suffolk.

- A species recovery programme in partnership with English Nature and the Suffolk Wildlife Trust for 20 of our country's threatened plants. One highlight, perennial knawel, only 4 native sites in the world.
- The village supports 133 scarce invertebrates and 21 Red-Data Book species.
- The nationally endangered robberfly, *Machimus arthriticus,* only two modern locations known in the UK.
- The RDB1 micromoth *Coleophora tricolor,* one of only four sites in Europe.
- Birds regularly seen on the village are goshawk, long-eared owl, nightjar and crossbill.

References

British Standard Technical Committee ES/1 (1996) *Environmental Management Systems – Specifications with Guidance for Use.* BSI, London.

Joint Nature Conservation Committee (1996) *Guidelines for the Selection of Biological SSSIs.* Peterborough.

Ruff, A. (1987) *Holland and the Ecological Landscapes.* Delft University Press, Netherlands.

UK Biodiversity Steering Group (1995) *Biodiversity: the UK Steering Group Report volume 2: Action Plans.* HMSO, London.

Index